My Heavens!

Other Titles in This Series

My Heavens!

The Adventures of a Lonely Stargazer
Building an Over-the-Top Observatory

Gordon Rogers

 Springer

Gordon Rogers
The Crendon Observatory
Near Oxford, England
01 degree 01 minute West
51 degrees 47 minutes North
Elevation 260 feet

ISBN 978-0-387-73781-2   ISBN 978-0-387-73783-6 (eBook)

DOI 10.100/978-0-387-73783-6

Library of Congress Control Number: 2007933700

9 8 7 6 5 4 3 2 1

Springer Science + Business Media

springer.com

To my beautiful wife Margaret who turns a house into a home, albeit with a gherkin attached.

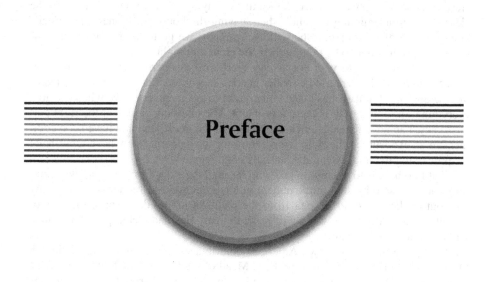

Preface

OK, so I am an "over-the-top" man. I spent $175,000 to build an observatory, before I even started to equip it, on an island country renowned for bad weather and too far North to see many of the deep-sky treasures. All this on a site that is pretty much surrounded by tall trees and where the subsoil is an unstable slushy sand.

Well, I always was a bit of a bull in a china shop: in fact, running the local cattle market, I once chased half a dozen escaped sheep into the carpeted Goldsmith's jewelry store. On another occasion my quarry was a 1-ton escapee bullock traversing the main railway line.

However, it was a chance sighting of a serious event on the Moon, seen only by me and those I brought to my small telescope, that put me on the path of acquiring the means to take photographs of what I could see in the sky. Now, I take pictures of many things up there that I cannot see—all this at a time when I was firmly of the view that these new-fangled computers were in the realm of youngsters and not for the likes of old wrinklies like me. (In fact, I think the technical word for me might be "mummified.")

I readily admit that technical know-how is something that I only slowly absorb. I do have business qualifications, which were hard won with excessive study over 5 years whilst some of my contemporaries just walked their examinations. Nevertheless, into the abyss I plunged with my first computer, with all its jargon and mysterious codes, and I thought I would be shot when I made my first "fatal error." Friends rallied round, though, correcting my mistakes and providing encouragement. Many wives would become disenchanted with their husband, missing at strategic times because there was a sky, but my gorgeous wife, Margaret, has been supportive throughout.

However, she has now developed the routine of saying, "Do not dare to mention the clear sky you are missing tonight" when leaving the house for dinner with friends after a month of solid cloud. She will even suggest we go to the movies tomorrow night because it is due to be raining then and clear that night.

I present this book to you, not as an expert with a superb grasp of all the finer points, but as a dedicated middle roader making mistakes galore but gradually learning the dos and don'ts. It has taken some years to get there, but now, when I switch everything on, only rarely is there a problem. The telescope points at the right part of the sky, I achieve focus, a quick calibration of the guide star, and have soon commenced an exposure at the right temperature for the camera. I now have a fair hand for processing the images, but the learning continues.

What I do here is to attempt to take you through the story from the ignoble beginnings and the step by step improvement in telescopes and techniques as I claw my way out of the chasm of despair in to the realm of taking deep space pictures of decent quality. Along the way I will provide many tips and techniques that I hope will be of help to the reader.

I have grateful thanks to two American contributors: Bob Antol of the Grimaldi Observatory in New York State and Brad Mead of the Spring Gulch Observatory in Jackson Hole, Wyoming. They both consulted me before embarking on their projects and very kindly wrote chapters about their domes and the equipment they have installed. Thoughts also for their very supportive wives. Thanks to Barb Antol for thinking of the fairy-tale idea of a circular row of portholes just beneath the dome, each with a candle light. Thanks also to Kate Mead for having the dexterity to drive a back hoe (mechanical digger) with Brad in the bucket atop a stepladder to high elevation in order to ascertain the line of sight over trees, without dumping him.

Heartfelt thanks also go to my son Chris for all the skill and dexterity he applied to building such a complicated structure and attaching it to a country cottage.

Also appreciation to John Watson at Springer UK for his encouragement, and for his skill at getting me back on track when I jumped the rails.

Figure 1. The author and his wife Margaret on the observatory Balcony

About the Author

Gordon Rogers is a member of the Royal Astronomical Society and of the British Astronomical Association. He is also a Chartered Surveyor specialising in large land development deals. This activity entails huge swathes of frustrating work with volumes of red tape. He tries to play golf; an immensely frustrating topic and remains doggedly on 28 handicap (the worst). Then he takes up astronomy, the mother of all ways to enjoy a frustrating time!

A chance sighting of an event at the moon ignited a slumbering interest in the wonders of the night sky and set in train the acquisition of a series of telescopes and observatories. These steps occured between 1994 and 2003 and culminated in the setup that is the subject of this book.

Struggling to handle all the leaps forward in technology he was very much a square peg in a round hole but gradually got there. His failings are legion but one talent he does have is diligence – he suffer all the disasters but gets there in the end.

Although having great resolve, he does not have enough to embark on supernova search in the manner of Tom Boles or Mark Armstrong. The burden of having to be up every night, and the tedium of visually cross checking thousands of images without success, would be a step too far. He therefore confines his efforts to taking pictures of known deep sky objects both within and without our galaxy.

He has taken untold numbers of astronomical images that have been failures for a plethora of reasons – jumpy sky – bad weather – bad picture framing – inadequate guide star – out of focus – inability to balance color – to name just a few. These do not linger in the memory but those that do were taken on nights of rock solid sky where all the set-up pre-requisites are good and the machinery performed impeccably.

His RAS membership opens the door to being authorised to give school talks. Numbering anywhere between thirty and three hundred pupils, usually in the age bracket of 8–11 years, he is always enthralled by the response from the youngsters to astronomy: they are fascinated by the subject. Invariably there are one or two "livewires" who will go on to be astro-physicists or something big in business.

Contents

CHAPTER ONE

Why Astronomy?

Maybe you are an experienced astronomer reading this out of curiosity, or perhaps you are a newcomer to the activity. If the latter, let me warn you about two aspects of imaging the sky at night. Firstly, you will need a strong constitution to overcome all the initial frustrations. A plethora of things will go wrong, and night after night you will wonder why you embarked, for pleasure, on such a tortuous path. Secondly, when you do eventually achieve an image of good quality from your own backyard and see that it bears a passing resemblance to something from *Hubble*, you will be gripped by a great excitement and a compulsion to get more pictures of better and better quality. This will have an effect on your social life and can cause you to arrange your calendar based on the lunar clock. You may find yourself at a dinner party sneaking glances at your watch in the hope that it will soon be over so you can get back to your beloved telescope for the first clear night in ages. It is an unwritten law that the only night of the week that is clear will be the one when you are booked to go out.

A word about me: at the outset I knew nothing of computers or electronics and now I am an experienced hand at astronomical imaging with a CCD camera. There are experts out there who leave me standing, but I do have perseverance. I relate more to the average guy struggling through a welter of information to be assimilated in the search for better quality images. Every day is a learning day. I do not have experience over a wide range of products and merely write concerning observatory structures, the telescopes, the hardware, and the software with which I have been involved. There is an awful lot out there about which I am ignorant, but I like to think that I tell the story of an average guy going about the business of trying to get pictures of deep space. With that said, let me embark now on the series of events that led me to build an over-the-top observatory.

OTA meant nothing to me until I was conscripted by His Majesty into the Royal Artillery for my 2 years of National Service in England. As a trainee Chartered Surveyor I was sent to Larkhill barracks on Salisbury Plain adjacent to the Stonehenge monument. I was to be an artillery surveyor and "flash spotter." It was here that I learned that OTA meant Optical Tube Assembly, or telescope, and received extensive training in the use of a theodolite for land survey. The other optical instrument was a pair of high-quality graduated binoculars that would be set up on a mount, orientated with north, thus giving the facility to call bearings to targets. After training, I was put on a ship to Hong Kong with the tannoy playing "You are on a slow boat to China." It was; the journey took 31 days. Many of us were violently seasick, but at least I established that the bromide had no effect. Fig. 1.1 shows the young author newly arrived in Hongkong.

One of my tasks was the calibration of guns by spotting the fall of shot at sea. For this purpose I was dispatched to the Straits of Malacca in Malaya during the insurgency. First, we had to survey the range, since there were no adequate maps. I have fond memories of setting up a theodolite, clad only in a bathing suit, in 18 inches of

Figure 1.1. A young Rogers on land survey for an artillery range in Hongkong.

warm water taking a mean reading as the hairline throbbed to the pulse of the waves. We worked over 5 miles to an accuracy of 15 inches.

On discharge from the army I went about completing my surveying course: this involved some more theodolite work, and I duly became qualified, but I did not at that time own either a telescope or binoculars. At great expense my father then sent me to an exclusive boys' school. Here, girls were not allowed across the threshold, but when I left school I did my best to catch up.

What has all this to do with astronomy? A date with a young lady would take you to an out-of-the-way pub for a romantic dinner and then a drive into the country for a talk. On one particular evening the air was balmy and the sky clear and moonless. To heighten the ambience further I opened the sunroof and laid back in anticipation of a good conversation. What I saw amazed me: the sky was so clear and deep that I was mesmerized, so much so that my date asked me if I was all right (I am not known for reverie.) I now realize that the sky that night was as good as it ever gets in the United Kingdom and that you may have to wait years to see one like it. It was more a sky from a good night in darkest Africa, and I was totally hooked.

I have always enjoyed photography and have owned a variety of cameras of reasonable quality. On the next occasion that the sky was reasonably clear I tried to take some photographs by placing a camera on a table in the garden and exposing the film for 5, 10, or 15 minutes. Although I got something, the results were an unmitigated disaster.

Gino and Peter were my hairdressers at the local Italian salon and provided me with regular updates on the activities of the local "petti mafioso" together with their "joke of the week." Next door was a local specialist camera shop where I became a regular customer, taking shots of people and landscapes. Calling in for some pictures, which they had developed for me, there was a sale and included in the sale was a telescope. No, not Tasco, but Zeiss. It was an ugly brute with bits and pieces sticking out in a most cumbersome fashion, but my camera knowledge told me that the optics had to be of high quality. The deal was done at half price. (Some years later I was talking to Peter one day and I asked him about his 15-year-old son Chris, whom I knew was mustard with computers. "Mr Microsoft has sent him another $1,500 this month."

"Whatever for?" I asked.

"Well, he has put together a Web site for Anna Kournikova and it has taken off: Chris gets paid for every hit on the advertisements." Now, my thought process might not be very quick, but it did not take me long to ask if the boy could put a Web site together for me.)

For the next few years the telescope would stand in the conservatory and get an airing, especially if there was a good solar system target. The Moon and the larger planets received regular attention and I had a particular high moment with Mars. It was in opposition and particularly close to Earth. Furthermore, it had good elevation from Oxford. I was astounded by the detail I could see with surface markings and a very well-defined polar icecap. A couple of weeks later I was watching Patrick Moore's fabulous *Sky at Night* program for the BBC when he showed a sketch of Mars that he had drawn using his 15″ reflector.

Wow! Through my ungainly little refractor I had seen all the detail that the Great Man had viewed. This encouraged me to step outside more often, and the next exciting experience concerned a moon of Saturn that would occult a star. The proximity of the planet made it a simple task to find the star, which was scheduled to be "extinguished" for 18 seconds, at a decent hour. As I watched it duly happened in full accordance, within a second or two, of the forecast issued a month or so before—what precision!

The evening of June 13, 1994, was the one that launched me on the astronomy path with a vengeance. I woke to a fabulous June day, the scent of the roses wafting in through my bedroom window. Breakfast was on the terrace, among the massed colors of the burgeoning plants all organized by the lovely Margaret. To be played at the local Golf Club that evening was nine holes of fun golf, followed by a barbecue. I had only a couple of beers and returned home to a warm and still evening with a Moon having a 30 degrees terminator. Let's have a look. Condensation. Well, I have never had that before!

Having wiped the optics these three white blobs persisted, so it had to be internal, although I had never ever before suffered with this. Since I was using an un-driven telescope, the Moon had drifted across the optics, and the "blobs" retained position relative to the Moon. This had to be cloud, but there was not one to be seen.

As it became darker, I could see more. Two of the "clouds" seemed free floating, but one was "connected" to a small crater on the moon, perhaps Santbech. From this crater issued a stem of milk chocolate color dissipating into the cloud. Each cloud had a width of around 100 miles, but the other two had no "connection" with the surface of the Moon. I convinced myself that I was watching an event on the Moon, so I rushed to get witnesses. My wife, Margaret, and neighbors Richard and Peggy Salmon with their daughter Hayley all looked and saw what I have described. I went indoors and did a crude sketch in my daybook of what we were seeing (Figure 1.2).

Now I needed to find a lunar astronomer or media person that may have had reports of this. My first attempt was a try to raise Patrick Moore. His number is freely available, but British Telephone inquiries could not find it. At Greenwich Observatory I got the janitor and had the same response at Cambridge University. I next made contact with the BBC, Independent Television and Oxford Radio. They all thought I was "on something."

In the United States I have two friends involved with television. Frank Rosa was then editor of TV news for New York City. Mike Marriott is a CBS newsman who made his name by filming the Vietnam War from the Vietcong trenches and married a Vietnamese woman. (He had promised her that he would not get into any further scrapes, and when he returned home wounded again, she threw pots and pans at him!) Both thought I was joking. I phoned NASA and got the answer phone. The moon was sinking, so my next four calls were to Hawaii. On the fourth—success! I made contact with a lunar astronomer, Kevin Polk, who had heard nothing but said he would look. Hawaii was clouded over that night.

Subsequent contacts with experts included:

Sir Patrick Moore. I had several most enjoyable hours and lunch with this amazing man who probably knows more about the Moon than anyone. He drew the maps for the Russian and *Apollo* landings, and with his non-stop observation of the Moon he forecast, through libration, the existence of the vast sea on the far side. He even named it *Mare Orientalis*. Patrick humored me, gave me a book of 700 TLPs (Transient Lunar Phenomena), all very minor compared to the grand event I claimed to have seen, and sent me packing.

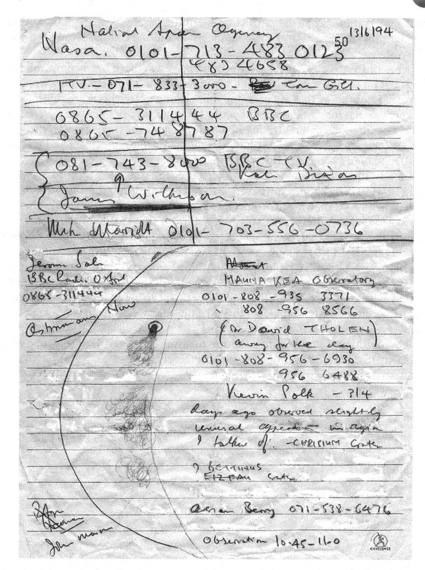

Figure 1.2. An extract from the author's daybook on the 13th June 1994.

Sir Bernard Lovell. In the meantime I received a letter response from Sir Bernard Lovell of Jodrell Bank Radio Astronomy fame. He thought I may well have seen an event but should refer to Patrick Moore as the ultimate expert.

Professor Spudis. At this time the *Clementine* mission was orbiting the Moon, sending back a wealth of detail, all of which was available to the public. Professor Spudis was running the mission and concluded that I may have seen exhaust gases from a spacecraft. If the spacecraft that was venting gases was at the Moon, it was pretty darn big, and if not at the Moon, then why did the clouds stay aligned with the Moon?

So what did we see? At the time we thought it may have been some volcanic activity, but on the following night with very similar conditions everything was back to normal. I now know that asteroids on approaching a large body are subjected to huge gravitational forces that cause them to break up, and perhaps we were seeing the effect of three fragments striking in a series such as the row of craters crossing Clavius. Whatever it was, I did not have the means to attach a camera and was too dimwitted to hold a camera to the eyepiece

Just a month later the comet Shoemaker-Levy was scheduled to impact Jupiter on July 16, 1994. The approach was very well documented, and there was a series of *Hubble* photographs showing the comet breaking up into fragments as it was affected by Jupiter's gravity. Although the strike would be on the far side of the planet, the thought was expressed that there might be a flash on impact. At Oxford it was twilight at impact time, but I had to look in case I saw a flash. Disappointment. Nothing. Of course, Jupiter rotates very rapidly, so I continued to observe and was stunned an hour or so later when a huge dark chasm came into view on the limb, followed by another. Each blow created a scar that was larger than Earth, and it seemed that an old friend had been seriously wounded. I invited a number of friends to take a look, and the universal comment was "very nice."

This event further reinforced my decision to acquire a good driven telescope with the capacity to capture images of what I saw in the sky.

CHAPTER TWO

Aperture Fever

8" Celestron

For the purchase of a new telescope I found a local supplier, David Hinds, who was the proprietor of the first proper telescope shop that I had seen. David advised that I purchase an 8" Celestron reflector, which was electrically driven. This was rather more complicated than the Zeiss Jenna, since for good results it had to be accurately leveled and polar aligned each time it was set up.

It is a known fact that if you buy a new telescope you guarantee a succession of overcast nights! Eventually the sky cleared, and with some considerable excitement I set up the scope on the terrace. Alignment with Polaris took some while but once achieved, I was delighted to see that the instrument would more or less follow the chosen object. Searching around the available bright objects it was rewarding to see them appear in the eyepiece, although invariably off-center since my alignment was not sufficiently accurate. So here we were with an operating telescope with the appropriate fittings to connect a camera body. Focusing was a bit hit and miss and, of course, there was bound to be the delay of having film processed before knowing whether you had it right.

Figure 2.1 shows one of my very first images. It is supposed to be M27, and you will see that it is not too good a representation of this nebula.

First, it is too bright, so it must be a different target, and second, the telescope is flaunting all over the place. The answer to all this movement was, I thought, that the telescope was set up on paving stones, and my pacing up and down for the 5-minute exposure worked wonders for its mobility.

Shortly after I acquired the Celestron, comet Hyakutake graced the Northern skies, and I had to have a crack at getting a picture.

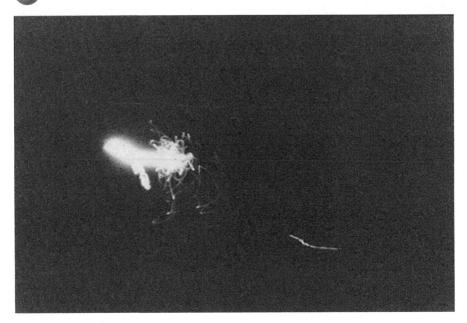

Figure 2.1. This early image on film is supposed to be Messier 27!

Figure 2.2. Comet Hyakutake on film 27th March 1996.

Figure 2.2 shows the result in black and white, although the color film showed a pleasing green fuzz that seemed the right color from what I read. It was out of focus, but at beginner level I was chuffed.

Not long after this there was the excitement of an eclipsed Moon at a decent time in the evening, when even neighbors popped out into their gardens for a look, since it had been well publicized. The Moon took on this gorgeous orange–copper color, and I took many exposures bracketing the duration widely in the hope of getting it right. With a bound I descended on Boots the drugstore for my pictures, to be greeted by the rather lovely young sales lady who had been admiring my handiwork and wanted to talk about it. Even if I say it myself a few of the pictures were spot-on, field of view, focus, and exposure just what they should be. A little success does wonders for the confidence, but all this was soon to be shattered!

Much was written in the astronomy press about film types and speed. ASA 100 had a superfine grain and ASA 3200 was superfast for faint distant galaxies. Some imagers understood the technique of hypersensitizing their film, but this was too much for me to attempt.

Acquiring a Computer and CCD Camera

It was time to move on and acquire this thing called a CCD that they were all talking about. Apparently it was an electronic gadget that would capture pixels of light and display them on a computer screen. Needless to say I had no computer and had been very much of the view that only the young could operate these new-fangled things. The first reluctant step was therefore to buy a computer.

I have always had the view that if you do not know the subject you find someone who does. I knew a man who understood. He was Martin, and he had a fantastic job as Commercial Director of Guinness Exports. Martin's business sent him around the world, and he was always coming back with stories from the Seychelles, Panama, Nigeria, and other far-flung parts. Can you imagine being paid to make sure that Guiness tasted OK? Yes, he would take me to PC World. All the jargon left me floundering, but I came out with a computer under my arm.

Now, my powers of memory are known to be somewhat fallible, and my grasp of what to do with the new machine took an age. Martin must have considered having his telephone line disconnected. Maybe my inept operation caused the fault to develop in the computer, but it was something Martin could not fix. Time to turn to the American boyfriend of a lady I have known for a while. This guy's most recent job was to sort out radar systems in the Middle East, and he was due to return to the United States shortly to install radar in an aircraft carrier. He was flummoxed, and we headed off to PC World. Apparently he had to know where the computer was built; with this information he was able to fix the problem. I was interested to watch him bamboozle the PC World technical guys!

Prior to buying the computer I had ordered a CCD from a manufacturer not far from Oxford, Starlight Xpress, and the call came through that it was ready. Could I get it to work with the computer? Needless to say, it was overcast, so I tried it indoors.

Figure 2.3. The author's first digital photograph.

Figure 2.3 shows my first-ever image with a CCD camera. I was amazed at the sensitivity of the chip, since the only light source was a glimmer from beneath a closed door, and the exposure was just 100th of a second. (My house is old, and some of the doors do not fit too well). The following day, cloudy, of course, I imaged a tree and achieved focus!

Eventually there was a sky and a Moon. I was transfixed with the quality of the image that could be acquired in seconds and took any number of images of craters, being careful to record their names and the dates of the observations. I started to see transient lunar phenomena—light plumes issuing from the Moon, I thought. No—they were just glints of sunlight, and I reluctantly proved this by imaging the same objects on a number of occasions. Gradually I was becoming more confident in operating the equipment, and I decided to have a crack at Deep Space. M51 was the first target, and Fig. 2.4 shows what I got. The signal to noise was atrocious, and the tracking way off. Figure 2.5 is a more recent image of this galaxy.

16″ Meade Altazimuth

At around this time I went to a meeting of my local astronomy society and had a chat with one of the sages after the meeting. We discussed telescopes. He advised that I really should have the largest one I could afford. The leading "big one" at this time was the 16″ Meade Schmidt Cassegrain, and he got me thinking. What a wonderful step forward, but where would I put it? My house has views that are heavily

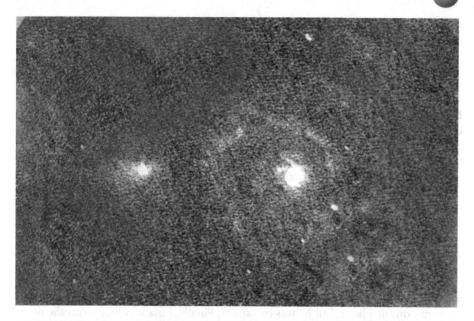

Figure 2.4. The author's first digital astronomical image (Messier 51).

Figure 2.5. A more recent image of Messier 51.

dominated by trees, so it would have to be on the elevated terrace, which has a modest line of sight to the south and west.

Would Margaret agree? Well, I eventually got her to agree that a small shed would be in order, but it must not be larger than 6 feet square. Off I went to the local garden shed man. When he heard my ideas about a shed with a roof that slid off I was surprised to hear that he had just converted one of his sheds for a customer (who happened to be a business competitor), and he would happily provide one for me. It came on a bright, sunny day, and the basic structure was soon in place. The roof was pitched with a gable end with doors that opened to pass the telescope. The transport system was a pair of garage door runners which I secured to the house and the whole system was easy to operate by hand and is depicted in Fig. 2.6.

The Meade could have an equatorial mount, but I had to settle for an altazimuth one because space would be so tight in the shed. Now, that was exciting when the guys came out from London to install it. Soon it was fired up and ready to go.

The altazimuth idea sounded fine. There is a bit of kit called a de-rotator, and this rotates once every 24 hours, eliminating the rotation of the scope. It did not seem to work for me, since every image showed circular trails.

After complaining for several months Meade sent me a new rotator, which also did not work. Then they sent me another one that did not work. I tried to reach the chief executive on the phone, but he was in Taiwan. Finally, I made contact with the man

Figure 2.6. The 16″ Meade in the shed observatory at the time of the nearly total eclipse of the sun on the 11th August 1999.

and gave him a blast. With many apologies he promised to personally deal with the problem and assured me that I would then have an "awesome" telescope. The fourth de-rotator duly arrived and, presto, when it was switched on, something different happened—it made noises. Nobody had bothered to program any of the three previous rotators that had been supplied to me!

Around this time I got word of a company called the Santa Barbara Instrument Group. They had made a camera that was very well received by the amateur astronomy society. It was an ST6, and it had large pixels for rapid light grasp. Why not?

I ordered the camera from London, and it took an age to come, plus the price, with all the taxes, was at the upper end of astronomical. Nevertheless, what a leap forward! The light grasp was exceptional. Messier objects could be imaged in just a few minutes and at a far quicker rate than previously. I thought I would try some more adventurous targets and chanced on NGC 4565, not expecting anything special. When the image downloaded I nearly fell off the stool as this flying saucer took over the screen. See Fig. 2.7.

I was sitting in a cramped observatory in the cold doing my thing. I thought, how about running some cables to the first floor landing that overlooked the observatory and creating a workstation there? There was plenty of room, but I had to drill through some pretty thick stone walls. The Meade has a big mirror, and I had an electrical Meade focuser but did not get on with it, so I focused by hand. Not so easy when the route from the computer to the observatory is along the landing, through the billiard room, down the spiral staircase, across the conservatory (being careful not to fall into the pond), out across the terrace to the scope. Getting focus could mean quite a

Figure 2.7. This image of NGC 4565 "knocked the author's socks off" when it downloaded.

number of these trips. After a week of good weather I certainly had no need to visit the gym.

The Meade was big and solid and did a good job. It also looked the part.

Then I had the grand idea. Why not become a qualified astrophysicist? There were advertisements for places for youngsters at Oxford University, which was just down the road, and I managed to find the name of the man involved. On the telephone Professor Charles said they did not usually take "outsiders," which implied that it might be possible, so I persisted. I thought I was making progress and then got shot down in flames with one question: "What is your grasp of mathematics?" End of story!

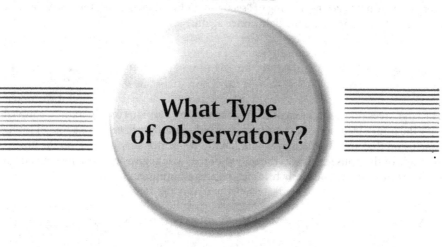

CHAPTER THREE

What Type of Observatory?

In my usual "over-the-top" manner I had by this time begun thinking about a proper observatory. Following were the five principal considerations.

Light Pollution

I am fortunate to live at the edge of the village of Long Crendon, sufficiently far away from Oxford, 13 miles distant, to be outside its glare. There are some local light sources, but by late evening they are usually doused. One particular problem was a streetlight to the south. Rob Gendler's difficulty in this direction was solved by throwing a blanket over the light every night, but I wanted something more permanent.

Hence, at 2.00 a.m. one morning, armed with a long pole with a paint brush attached, I daubed my side of the offending light liberally with black paint. Not long after this the bulb failed, and the road people got the message and decided to leave it in this state. (Amazingly, the local council has now adopted a strong anti-light pollution policy. The lights have been changed to shine only downwards.)

One neighbor's 500 watt "security" light was particularly troublesome, and I erected a board to screen it. On seeing this he kindly got the message and volunteered to move the light to a different wall. About 4 miles away due south was a floodlit depot with immensely powerful lights that killed all those gorgeous southern targets. A word with the site agent, and he was happy to just tilt the beams a little towards the ground and solve the problem. These difficulties aside, I have a good sky and often can quickly see down to magnitude 5. While the many mature trees restrict the available sky, they can obscure light pollution when in leaf, and some are coniferous. (British Astronomical Society member Gary Poyner is a most accomplished Variable Star Observer from a badly polluted site in Birmingham. He is also a cat fancier and feeds the strays and others. Hearing him going outside to observe they run in from every direction, triggering a string of "security" lights. Replete, they then dawdle back to their stations, triggering the lights once again.)

Topography

The large trees on all sides except the northeast were a serious consideration, coupled with the quite steep slope of the site from northeast to southwest I was forced to the conclusion that any observatory would have to be located very high up on the northwest side of the house. This was the only spot that would give access to the many deep space targets that are southerly from 51 degrees 47 minutes North.

Type of Enclosure

This is a knotty one! Should it be a slide-off roof, a clamshell, folding panels, a dome, or something else? I think, at ground level, I would have chosen a slide-off roof. Once opened, the whole sky is yours, air stability is at its best, and the observatory can be closed manually in the event of a power cut. On the other hand, a dome will give shelter from wind and light pollution and give some protection from the rogue shower of rain. I read a lot about available observatories and went to see some. Let me relate a story pertaining to this.

I was staying with friends in Maryland and knew that a manufacturer of fibre-glass domes was nearby. We were going to Greenbelt for a Hubble ST tour, and as it was on the way, we called in. "By all means come and have a look." I found what they had to be an efficient and economically priced unit, and the proprietor then suggested we take a look at the sun. Explaining the dangers involved to my friend David, he then projected the beam onto a board, but in so doing, in a split second, the beam accidentally crossed a wire. There was a flash of smoke, and the wire was laid bare in a moment: David got the message! On the Greenbelt tour, the guide took us to the door of the room from which Hubble is operated. I asked if he knew what they were imaging. He did not and volunteered to inquire; after a couple of minutes he emerged, shaking his head, and said "They do not know." Presumably they have orders to take exposures with various cameras at various coordinates and that is that.

A couple of years later I met David again for the day and he asked what would I like to do? I told him I wanted to go to the Space Telescope Science Institute. So off we go to Hopkins University, but we cannot find the place. Everyone we asked assured us we are in the right vicinity, but no one knows the exact whereabouts. I made enquiries at an astronomy museum, and they directed me to the rather modest structure at the bottom of the slope on the opposite side of the road. I could see no signs proclaiming the wonderful work that is done there. The exceedingly attractive receptionist greeted me with a big smile when I explained that I was an astronomer from England, and it would be a huge bonus for me if I could talk to one of her professionals. She could not have been more helpful, but it seemed everyone was out, engaged, or on the phone.

After 5 minutes of efforts she advised me that she had found someone who would be down in a few minutes.

Sure enough, down came this guy who handed me his card—Mario Livio, Head of Science. She had landed the big one for me! Needless to say it was fascinating to look around the building and down in the basement, to look at the computer room where all that mind-blowing information arrives, with just two people in attendance. Acknowledging that he was a very busy man I said "I have just one question: What is the most important thing you have been involved with concerning Hubble?"

"It has to be establishing that the cosmos is expanding and accelerating in that expansion."

Available Ground

My considerations led me to the conclusion that the only possible site would entail the purchase of a small strip of unused ground owned by a close friend and neighbor, Richard (he witnessed the events on the Moon in 1994). This sounds easy but, hang on, Richard is as generous a "mine host" as you will find when he is at home, but in business he is a tough dealer, and this was business. After months of negotiations I owned the land, and Richard was building a new wing on his villa in Spain.

Will Margaret Agree?

I think I can safely say that, generally speaking, ladies do not see astronomical domes as things of beauty. In addition they cost money that could be spent in other ways. Margaret wanted a new kitchen. There is nothing wrong with our kitchen; she designed it herself, but she fancies a change of color. Now here I must relate a masterstroke of negotiation. We were staying with friends in Portugal. They had a new kitchen, which Margaret much admired, and she took a special interest in the knobs of a local pottery. Would you believe that at a cost of 48 Euros I bought 48 of these knobs, which I changed for the old ones as soon as we returned home. Agreement secured, I thought, but she also got a new car (although I did not have much choice in the matter, since while her car was being driven to the garage for service, by the garage's driver, it spontaneously burst into flames. The fire Brigade said it was the best car fire they attended all year long!).

Decision

I now knew where I will build the observatory and that the floor of the dome would be some 20 feet above ground. Being on rising ground, the site is swept by fierce southwesterly gales from time to time. This made me conclude that strength was the prime consideration and caused me to settle on a dome as the preferred type and a steel one for strength. The astronomers at the University of Hertfordshire's Bayfordbury

Observatory recommended Ash Domes to me, and this was the path I chose to follow. It was very helpful to be able to see their domes and take measurements; they even had a 16″ Meade in a 10′6″ Ash dome—just what I had in mind. I also went to London University Observatory at Mill Hill. They have Ash domes, and the professionals there gave me very sound advice, too. At the end of the visit I was asked if I would like to see the workshop. Down in the basement was Fred in his brown smock. On the workbench was a rather cumbersome instrument upon which he was working. What was it? The spectograph for the Keck in Hawaii. Why was he repairing it? Because he made the darn thing in the first place!

If Ash was good enough for these two observatories, it was certainly good enough for me, so the decision was made.

CHAPTER FOUR

Town and County Planning

Background

My father was an agricultural auctioneer, and I went into the business on leaving school. They let me sell the day-old chicks and calves and occasionally the sheep. Later I was put in charge of an auction room selling secondhand furniture that had seen better days. From time-to-time, there would be an item of high quality, and I would call in Frank and John Bly of TV *Antiques Roadshow* fame to give me an opinion. I hankered for something both more challenging and more rewarding. This proved to be negotiating the sale and development of land.

It all started from a footpath diversion order over allotment gardens that resulted in a fee of 7 guineas for 6 months work! This gave me the opportunity to ask the owners of the allotments if they had thought of building on the unused part. They had considered it but had done nothing about it, so off I went to investigate with the planning authority. This was the first of many deals on the ever-more complex path of securing consent to do something with a piece of land.

To cement my credentials I had to become qualified as a Chartered Surveyor. This entailed an arduous 5-year course of postal study when, for 9 months a year, five evenings a week, after a full day's work I would undergo my correspondence course. This meant turning down my pals' frequent invitations to go down to the pub for a quick beer. At the end of it, I had the magic qualification "Associate of the Royal Institution of Chartered Surveyors," and that gave me a stature that opened doors in the business world. Once qualified I found that I was doing more and more work of a planning nature and was even getting better results than some eminent planning barristers. I knew the system and I knew the officers and what they were looking for. (My last deal took 16 years from presentation until building work commenced. I called to see the County Surveyor and said, "You need a road to connect the Oxford Road with the Bicester Road, but you cannot afford to build it."

"Correct," he replied. I then explained that a developer would build it if consent were given for development, which would be a fairly obvious rounding off of the Town of Aylesbury. He agreed, and all those years later, after 'umpteen meetings in chambers in Lincoln's Inn to consider all the documents, permission was given for 2,500 houses and 60 acres of commercial development.)

Local Support

My knowledge told me that getting consent for an observatory would be a delicate operation, and that local support would be critical. This was 1999, and what I was proposing was a true Millenium Dome. Nearby in the village there was an ongoing dispute where a phone operator was looking to erect a mobile phone mast close to the village school. It was proposed that this would cause all manner of defects in the children, and the local member of Parliament was up in arms and leading the opposition to the mast. The erection of this mast was a new high-tech operation, and my observatory would be seen in the same light. One bit of luck was that my house was just outside the conservation area, but it had a thatched roof of reed that came all the way from the Camargue in the south of France, so that an adjoining high-level dome would be seen as too outlandish.

Located as far as you can get from the sea in England, the village has a very quaint old High Street with a jumble of houses from different eras and is fortunate in that it is now a lightly trafficked by-road. In Coronation Year 1953, Long Crendon was named "Typical Old English Village," and it is supposed to contain the oldest house in the country, although however would you know that? It even has two manor houses. One, dating to 1180, is frequently used for television filming for the likes of Midsomer Murders. I know the village better than most, having lived in four houses there with three different ladies.

Long Crendon is indeed a long village, spreading well over a mile in length, 280 feet above sea level, and with 2,500 inhabitants. It is very much sought after, with many enchanting corners but it also has its blander, more recently built parts. There is a post office, which is something of a village hub, where you always get a friendly welcome, a butcher shop, a news agency, stores, and a garage. There is an award-winning English restaurant and three Oriental restaurants, a village Club, Tennis Club, Bowls Club, Cricket Club, Pigeon Racing Club, and Football Clubs for seniors and juniors.

There are several inns. On a summer's evening they are sometimes frequented by Morris Dancers, who go about their Old World form of dance with bells on their feet in costumes of a bygone age. Some of the houses have glorious gardens that are opened to the general public twice a year. The village is skirted by the river Thame, which is a tributary of the Thames. At times of high rainfall, the valley of the river Thame is deliberately flooded to alleviate problems in London.

Like much of the world our weather has changed in recent years. The last 12 months were no exception, as a summer drought was followed by endless winter rainfall. For astronomy not much has changed. We have good weather but not very often, although the pattern seems to be evolving into longer good periods mixed with more persistent bad times.

The planning rules in my area are that if no resident objects to a minor application the Planning Officer can determine the case. I conjectured that if all the local residents wrote in support, he would have no reason to rock the boat. My first move, therefore, was to write to all the near neighbors, sending them plans of what I was thinking of and asking if they could support it.

Amazingly, despite the raucous parties, I had not upset any of them sufficiently to prevent them from lending support and writing in to the authorities saying so. However, I had made a mistake. I had not cast the net far enough and had failed to contact some distant occupiers who were above me to the East and partially screened from what I was doing. Two objected, including one couple with whom we regularly shared drinks at parties! On the warpath I went and descended on each objector to see if I could get them to relent. The "friendly" couple could not see the site of the observatory from their house, but they might get a glimpse from a seat in the garden. After having several of their whiskies I managed to persuade them to withdraw on the condition that I would plant a mature bush and draw down some branches from laurel trees in my garden to screen the offending structure from their vista from the magic seat. The other objector would have a clearer view, and I got no refreshments from my talk with him. If I could plant a mature tree on the line of sight he might withdraw, but I did not own the adjoining land, although I knew who did—a tax exile in Jersey. A telephone call to Jersey and another meeting with the objector and terms were agreed for me to plant a mature sycamore.

The local Parish Council had a voice in planning matters but not a strong one. I learned that while debating my application concern was expressed that no bedroom window in the village would be out of my view. On this topic I should relate that, later, the son of a neighbor and good friend was courting a gorgeous golden disc pop star, and the room in which they slept directly faced the observatory at a distance of around 150 feet. Hearing about the lovely damsel a host of friends suddenly wanted to descend and do some observing. They were incredulous when I told them the scope would not focus under around a quarter of a mile.

Obtaining Consent

With no current objections and many letters of support the planning officer put through the application "on the nod," and it escaped the scrutiny of the District Council planning committee.

The consent came with the usual conditions but there was one specific item: the dome was supplied in a silver finish, which was a good match for the clouds of England, but the dome was required to be painted a copper color (more about this later).

Aligned to planning in England are the Building Regulations, application of which is also operated by the Local Council. The Building Inspector ensures that buildings are soundly constructed, and he has a rulebook, which is rigorously applied. Because of the very nature of an observatory access from below is invariably complex. After much consideration my son came up with a masterful plan that would give a proper staircase entry to the dome; however, this was dismissed out of hand since 6 feet of headroom was inadequate! Result—access was to be by ladder—progress!

CHAPTER FIVE

Construction

Finding a Builder

I had now acquired the necessary land, had secured planning consent and needed to find a builder. I required someone who, on dodgy ground, could build a three-story structure and attach it to a house, which was already much extended. There would be all manner of difficult abutments, with much leadwork and the interleaving of many tiles. Where the dome cut through the roof, lead soakers would have to be individually measured and cut, and the flat roof balcony presented a particular difficulty. The whole thing gave numerous points of concern for water penetration. If I went to Yellow Pages seeking a builder, I had no doubt that "Get lost!" would have been a pretty universal reply.

However I have a son, Christopher, who is a jolly sight smarter than me. As a lad he won the "Silver Trowel" for best apprentice bricklayer in the south of England and went on to be top apprentice overall. He built his own house, living on the site in a caravan, and gradually progressed. By 1988 he was stretching his wings when the housing market collapsed and he was very badly exposed. While his peers were going bust all around him, Chris fought and fought. He negotiated his way out of some nasty positions, and he took on, and became expert at, the only field where there was plenty of work—renewing defective foundations in the many houses that were suffering from subsidence in west London, brought on by drought. The attraction of this was that he knew where the money was coming from, as it was at the behest of insurance companies. Having just kept his head above water he moved from contracting to developing and made quick progress as the economy recovered. Now his father was looking for a contractor for a difficult build. Chris volunteered to take it on.

Groundworks

The house was located on rising ground, elevating it above the Vale of Aylesbury. The clay vale is renowned fatstock cattle country, and Aylesbury town is renowned for its ducks. (Many husbands, in endearing terms, refer to their wives as "my old duck.") In the summer the sky above my garden is a popular spot for gliders seeking thermals overhead—probably not a good sign for air stability. The substrata are extremely varied, consisting of limestone, clay, sand, and even some chalk. Camellias will happily grow next to Campanulas, although the deer tend to eat the former!

At the spot where the observatory was to be positioned I knew there was a potential problem for two reasons: (a) In preparation for some yard resurfacing I was merrily digging away when the ground subsided and I disappeared up to the neck into a sandy pit. This I now found was the point where there had been a well. (b) On building an extension to the house some years before we encountered a particularly bad piece of ground at a corner of the structure. Going as deep as we could we could not find a "bottom," just a continual mushy sandy-clayey "goo." It took the structural engineer some days to design the reinforced foundation to overcome the problem. Clearly we needed expert help if we were to stand a 20-foot high concrete pier at this spot! Yes, said the specialists, first we need to thrust bore down as far as necessary. This entailed thumping a probe into the ground and taking a sample for analysis every 6 inches. They did not like what they found and carried on down to 33 feet. (The several 12-pound Koi carp in the adjacent indoor pond must have had a headache from all the banging.) Off the samples went for testing, and a couple of weeks later the specification came back. The pier should be supported on three piles, each 9 inches wide and 27 feet deep. Still no "bottom" had been detected, but the friction of the piles would sustain the pier. To the consternation of the neighbors the drilling truck arrived (Fig. 5.1). It was huge and carried all sorts of paraphernalia.

The plan was that, on drilling a hole, reinforcing rods would be inserted and the hole filled with sloppy cement, since concrete would not be sufficiently viscous to fill the void.

Figures 5.2 and 5.3 show the thrust boring and drilling. At the end of the day the piers were in place and a reinforced cap was cast over them with rods in place ready to tie into the pier.

Why would the sink not empty? asks Margaret. Oh dear! The drain is not working and yes, you can guess the answer: we have punctured it.

Figure 5.4 shows the work of reparation in building a diverted drain.

Superstructure

So now Chris could get on with the building. Surprisingly we found a good bottom for the foundations to the walls and construction of the carcass proceeded apace. Figures 5.5–5.7 show some of the early work.

It was decided that the walls at the upper level were to be timber framed, since I was anxious that the structure dissipate the heat of the day as rapidly as possible. Within a framework of blockwork piers the wooden carcass was constructed, lined internally with insulating felt, and clad externally with timber, as agreed with the planning authority, to copy the appearance of old agricultural barns to be found in the

Figure 5.1. Consternation in the village as the drilling truck arrives.

Figure 5.2. Thrust boring to 36 feet for soil samples.

Figure 5.3. Drilling to 27 feet for the piles to support the pier.

Figure 5.4. Repairing the drains having drilled through them.

Figure 5.5. Foundation time.

Figure 5.6. The pier base.

Figure 5.7. Starting the block work.

surrounding countryside. At this time we had an elderly cleaner called Ruby. She was a dear soul and worked to a fixed schedule, which included opening the back door and shaking the dusters at around 11.30 a.m. The chaps enjoyed her consternation when, upon opening the door, she found it to have been bricked up.

A void was left for the pier since we were unsure how to best contain the concrete. The structural engineer thought 18 inches width would be sufficient, but in my usual "over-the-top" manner I opted for 30 inches. Chris eventually found a length of heavy duty ribbed plastic sewer pipe that was strong enough for the job (coincidentally Bob Antol in Chapter 14 arrived at the same choice). He calculated that it would take 8 tons of concrete to fill it; of much concern to us was the fear that the weight of the mix might make it explode from the base. Margaret would not be happy with 8 tons of concrete outside her kitchen window. Having gotten the tube in place we decided to build a retaining wall around the base and to leave it several days to cure. Then came pouring day, and Steve stood by his mixer with the ballast and cement. (Steve would shovel 8 tons that day.) At a third full Chris called a tea break and took a sample bucket of concrete. Not until this had started to stiffen did he carry on with stage two and then the final stage after a suitable pause. Figures 5.8–5.14 show various steps in the process from the arrival of the tube onwards.

During every step of installing, the pier we took great care to insulate it from the surrounding structure. At ground level there was a good clearance from the paving and nowhere did the fabric of the building get closer than an inch or so. Voids between

Figure 5.8. The tube to contain the pier concrete.

Figure 5.9. The rolled steel joists.

Figure 5.10. Carcassing work in progress.

Figure 5.11. Pier reinforcement.

Figure 5.12. The pier tube in place.

Figure 5.13. Internal carcass.

Figure 5.14. Topping out the pier.

the pier and floors were filled with fiberglass wadding or carpet. We were fastidious about this. (With the telescope sheltered from a good breeze by the dome everything would be stable. However, when it really blows there is vibration at the telescope in tune with the wind. I think as the blasts of air hit the house they cause the ground to vibrate, and there is no better seismometer than a telescope looking at infinity.)

The Ash Dome

Having decided on the manufacturer I now had to settle the specification. The minimum dimension across the base was 10 feet 6 inches. I would have liked a larger dome, but there was not room. From my tour at Bayfordbury I knew that this would accommodate a 16" Meade because this was the setup there.

The dome could have a standard shutter or an extra wide one: I opted for the latter. This I found to be a good move, since it gave a slot of about 2 hours. When imaging piggybacked the decision was especially useful. At some points in the sky the telescopes are parallel with the horizon, and this effectively halves the width of the opening. At other points when running more vertically the slot can last for hours. Now the bad decision: Ash offers an option of a clear lanphier shutter that can be attached to a draw box, giving a totally enclosed window on the sky at any position.

I thought this would be a good idea for displaying the glories of the sky to visitors, but I never had in mind to image through glass. Visitors come and go, but I have to live with the shutter full time. As a target rises or falls, this can entail moving the shutter by either parking it or raising it; you will only be doing this at a time of good imaging sky! The basic dome comes with a bottom section, which hinges outwards so that, once opened, the whole slot is yours for the night.

Having made my selections the order was placed, and it would take about 3 months to deliver. This fitted in very well with the building schedule, and, on time, I received the message that the dome was about to be dispatched. Not only this, but I was e-mailed a photograph of the assembled dome at the works outside Chicago, with my name on it. This was a clever ruse by Ash. The message was, "If it does not fit together, the fault lies with you, the buyer."

Delivery was to be to at the Southampton docks, and the shipping cost was a separate item in my Ash bill. It was for me to get the container from Southampton to Long Crendon, a distance of around 80 miles. Needless to say this cost more than the Chicago to Southampton stint.

The big day dawned, and since the truck cannot get anywhere near my house I arranged a rendezvous in a nearby lane (Fig. 5.15).

You will see from the white painted stones that the adjoining owner was prissy about his grass: the result was a tirade of abuse, which was enjoined by another resident. We proceeded to open the container.

Figure 5.15. The container lorry arrives.

Figure 5.16. Terry is somewhat daunted by the contents of the container!

What a sight greeted us! Terry was Chris's right-hand man; he could turn his hand to most things, but was this a step too far? (Fig. 5.16). The first surprise was that, apart from the shutter, all the components were small. There was a melee of bits and pieces, steelwork in all shapes and sizes, and unintelligible timber parts. There were boxes of nuts, washers, shims, and various other things. This all came with a 76-page book of assembly instruction, an extract from which appears as Figs. 5.17 and 5.18.

My first thought was, "What have I done?"

The first step was to try and identify the various parts; we made some progress but some bits baffled us. (At the end of the job a couple of small parts were left over. Was this someone at Ash having a joke?) The instructions proved to be splendid. Each step was clearly identified with copious drawings that clearly demonstrated what had to be done. As the building was of timber, I had arranged for a local steel fabricator to make a steel ring beam for the dome walls to sit upon (Fig. 5.19).

We fully understood why the instructions emphasized the need for everything to be totally level. The base timbers of the dome walls married up to the ring beam, so we had got that right. Next we assembled the plywood walls to the dome (Fig. 5.20).

Now, the fundamental move of installing the dome runners: the more accurate the leveling, the better the dome would operate. Chris is very good at attention to detail, and he got this spot-on (nowhere in its trundle does the dome labor) (Figs. 5.21 and 5.22).

During the construction there were a couple of occasions when we needed some clarification. An e-mail query to Ash always resulted in a rapid reply with a solution.

ASH-DOME | MODEL "REA" BASIC STRUCTURE | R 101

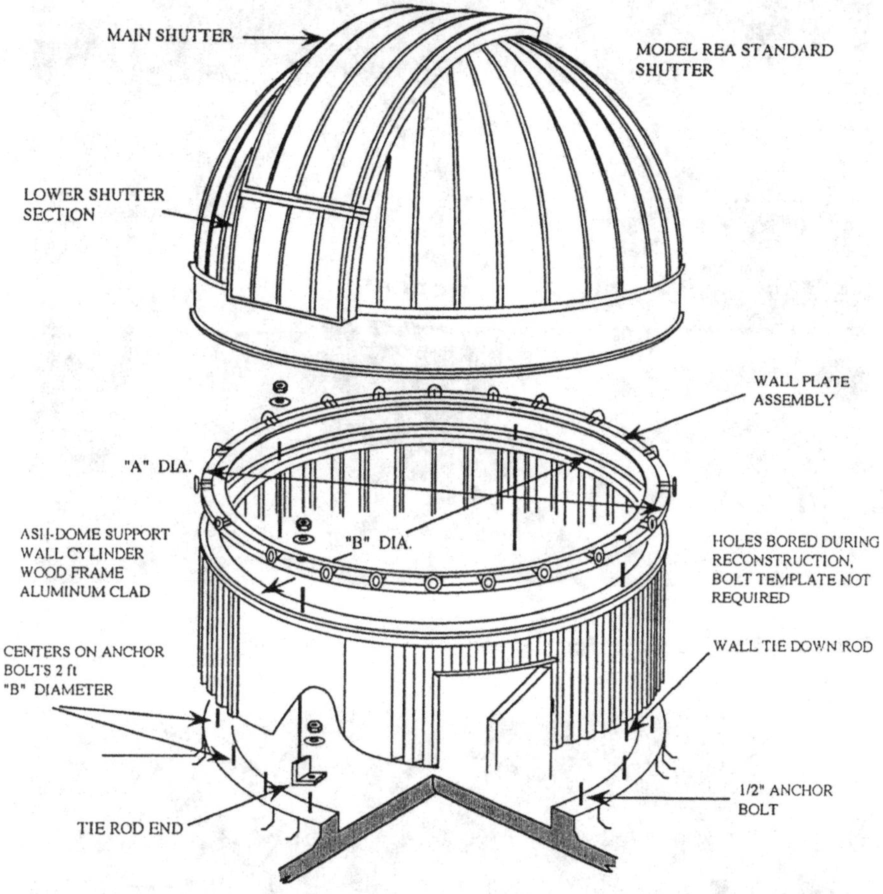

DOME DIAMETER	"A" DIAMETER OUT SIDE WALL PLATE	"B" DIAMETER ANCHOR BOLT CIRCLE	ANCHOR BOLTS REQUIRED	ANCHOR BOLT SPACING FOR WALL PLATE
8'0" (2.43M)	7'5" (2.26M)	6'11" (2.10M)	15	17-5/8"
10'6" (3.20M)	10'2" (3.09M)	9'6" (2.89M)	20	17-13/16
12'6" (3.81M)	12'2" (3.70M)	11'6" (3.50M)	20	21-9/16"
14'6" (4.42M)	14'2" (4.31M)	13'6" (4.11M)	24	21-1/8"
16'6" (5.02M)	16'2" (4.92M)	15'6" (4.72M)	28	20-13/16"

Figure 5.17. An extract from Ash's instruction manual.

There is a footpath crossing the field adjoining my house, and, as the observatory started sprouting in the air this caused us to get some funny looks (the path is now known in the village as "Observatory Walk"). This circular plywood upstand looked very "raw," but as the metal cladding was installed it adopted a more businesslike

Figure 5.18. The shutter "on the road".

Figure 5.19. Positioning the steel ring beam for the observatory.

Figure 5.20. Assembly of the dome wall.

Figure 5.21. Positioning the dome runners.

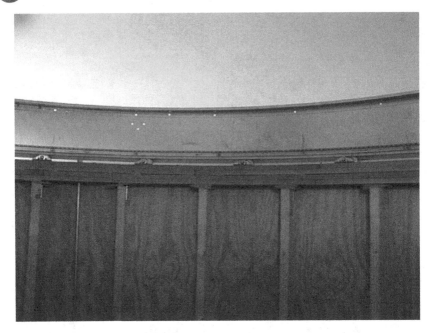

Figure 5.22. The dome runners and base panel.

Figure 5.23. Not your normal house extension.

appearance (Fig. 5.23). I liked the fact that Ash supplied an extra piece of cladding in case you spoiled one. We were now at the stage of completing the waterproofing arrangements. The tiled roof had to be fabricated and tiles and leadwork individually cut to suit the circular dome walls. I considered it essential to have access to a balcony at dome level (Figs. 5.24 and 5.25). Not only would it give a good means to check incoming weather, but being high and dark, it would offer a splendid vantage point for viewing the heavens with the naked eye and with binoculars. The roofline was carried up to safeguard the balcony on two sides, and the third side had to be co-joined to an existing tiled roof in a most complex fashion.

The only mistake made was the way in which rainwater was to be discharged out through the roof. This was simply by means of a lead-lined unenclosed opening in the studwork, and it did not occur to us to think about animals. Some months later, I heard noises in the roof and yes, we had rats! We straightaway closed all possible access points and put down poison. The rats eventually succumbed, and I was fortunate that the remains were all within reach of the trap I had made for the purpose of planting poison.

Now the exciting bit: dome assembly, and Terry missed it. He reported in sick, but we thought his illness might be related to an extended visit to the pub on pay night. The dome panels were made of aluminized steel, and the idea was that you slid them together with the aid of some dishwashing fluid (Figs. 5.26, 5.27 and 5.28).

Figure 5.24. The high level observation deck.

Figure 5.25. Tiled roofing.

Figure 5.26. One of the dome sections.

Figure 5.27. Starting assembly of the dome.

Figure 5.28. The dome panels slide together with surprising ease.

It all sounded a bit hit and miss, but, with firm pressure, they knitted together surprisingly easily. In only a couple of hours or so the dome was completed on a gorgeous autumn day, and we stood back to admire the scene (Margaret was not so enamored).

Arriving the following day Terry had missed all the glory and swallowed hook, line, and sinker the story that two gorgeous young Russian ladies had popped in to see us, sent by Oxford University. (They actually do this to me sometimes!)

Now came the most difficult operation of the whole exercise: installation of the shutter. Since I had ordered the oversize opening it followed that I had an oversize shutter. There is actually a crane depot at the other end of the village. Somehow the driver managed to squeeze into the garden and the process commenced. It was a squally day, and the lads were reaching out for the shutter as the crane driver offered it. A gust of wind sent it spinning like a top, and I feared for the boys' hands (Fig. 5.29).

All praise to the crane driver, who in no time took it out of range. It was sometime later before things had calmed down sufficiently to have another try. Slotting the shutter into position and securing it had to be the most difficult part of the whole operation. Of course, at ground level it would have been so much simpler. After that was done the crane driver maneuvered the box containing the roller shutter, which sits at the foot of the shutter opening. Next followed the installation of the motors driving the rotation of the dome and the lifting of the shutter (Figs. 5.30, 5.31, 5.32, 5.33).

Figure 5.29. Installing the shutter.

Figure 5.30. The rotation motor.

Figure 5.31. The shutter and its motor.

Figure 5.32. The curtain box attaching to the lanphier window.

With these in place the dome was completely fabricated. Another black mark for the lanphier is that the box for the blind has a nice right angle corner, which is exactly what you bang your head on in the middle of the night when exiting to the balcony to check the sky.

Electrical Work

We have an electrician in the village, Colin, who knows what he is doing. I told myself not to be caught out by being short of electrical points. I had three lighting points (not enough) and six power points (not enough). I have not automated the dome rotation, preferring to operate this with manual switches. The mechanism of the shutter worked brilliantly and the motor automatically shuts off at the very moment when the clasp is triggered by contact. Being on rising ground the area is subject to a good number of lightning strikes, and, of course, domes are tasty targets. A lightning conductor was installed on each side of the dome, but so far I have been lucky. Not so the UK's leading supernova sleuth, Tom Boles. He had the dreadfully bad luck to have his local telephone exchange struck by lightning, and the surge carried down to his computers and three telescopes, causing serious damage.

Figure 5.33. The shutter release mechanism for the lanphier window.

Telescope Mount

Figure 5.34 shows the pier head with three wooden inserts, to be removed when the concrete was set, so that studs could be accurately aligned for the equatorial pier, which was raked at 51 degrees 47 minutes. The circular device in the picture is really cool. The theory is that the pier for the LX200 is rotated by brute force and elevated by means of shims. A genius by the name of Alex Colburn designed and had made this adjustment device. So precisely was it engineered that I could rotate the whole telescopic structure of quite some hundredweights with my fingers alone.

Figure 5.34. The head of the pier and adjustment device.

Painting

My planning consent required that the dome be painted a copper color, so off I went to my friendly paint man. He could supply heat-reflecting paint in white and in silver but not copper. On the Internet I found a source at Biggin Hill, the old World War 2 aerodrome. They were very helpful but not cheap. Apparently they spend most of their time painting planes and missile silos! The paint would be custom-made, cost around $30 a pint, and the minimum order was around twice the amount I needed for my job. It would take two men several days to complete the technically challenging job of preparing the metal, etching, undercoating, and top coating the paint. I had to find local accommodation for them while they carried out the work and also had to arrange for scaffolding to be erected. Naturally, the weather did not cooperate, and I spent hours looking at weather charts. The final result, though, was superb (Figs. 5.35 and 5.36).

Figure 5.35. The dome prior to painting.

Figure 5.36. The re-painted dome.

Telescope Installation

At last the really big moment—the installation of the telescope. With plenty of hands on the ropes and some deft negotiation of the ladder, we quickly had the equatorial instrument in place, so that completed the physical work. Now all I had to do was align the scope with celestial north.

In fixing the position for the pier I had taken multiple readings of the position of the mid-day sun, adjusted for my position 1 degree 1 minute, west of Greenwich (Have you been there with your loved one and had a photograph taken as you embrace while each is standing in a different hemisphere?) I had also set up a ranging rod at the most distant point I could see to the north, settling its position by the Pole Star and checking with a compass. You will not be surprised to learn that it continued cloudy, and there followed a fearful storm. Huge blasts of wind and torrential rain that was nearly horizontal. Inwardly, because of the very awkward nature of the many abutments, I thought I would settle for a couple of points of water ingress. This was a storm from the south-west, the direction of the strongest winds we get in these parts. Search as exhaustively as I might the only water incursion I could find was a few drops from the shutter. From the structure, nothing! A little extra filling to the shutter solved this leak.

When the weather finally broke I was relieved to find I was not far out in alignment north and south and very comfortably within the adjustment capacity of the pier head. The finished dome, before painting, is shown here (Fig. 5.36).

One design feature we were able to include was a lobby to act as an airlock when proceeding from the computer room to the dome. In the lobby hang the duffle coat, Cossack hat, and ice-workers boots: all fundamental items when working on alignment or a technical problem in mid-winter. Umpteen cables have to run from computer room to observatory, and I got this system right the first time. I bought some 2″ plastic boxing and planted this on the surface of the walls and ceiling to form a channel that could be opened and closed at will. I have been in and out of that boxing on quite a few occasions. Something else I did for the LX200 was to arrange for the insertion of a 6″ drainpipe, with capping screw, in the floor of the dome, making it possible to drop the hand controller to the floor below.

The Computer Room

Having spent a fortune on building the observatory, my idea was to have a really cheap flat-pack but functional computer room. Oh no, Margaret would not go along with that. This would be part of the house that people would see, so it was off to the designer for a custom-made deluxe office suite. (She does not know, and would have a fit if she did, that two or three times a year there is a hatching of flies and lacewings in the observatory. I think they may be attracted by the copious amounts of grease on the rotation gear. The solution is simple: when she is out shopping I nip up with a vacuum cleaner and quickly dispose of them from the lanphier window, where they naturally accumulate. A more welcome visitor one day was a Goldcrest, a bird I had never seen in the garden.) Once again I told myself to have plenty of power points. The six that I allowed fell far short of what was needed, and the working surface area had become a little cramped with the advent of larger monitors. On the plus side the aforementioned Alex Colby did provide me with a wonderful gadget for fine tuning the pointing of the telescope: the little control knob sits close to one of the computers.

One thing I got wrong was to fail to clearly label every single cable that was installed. There is such a jumble now under the desk: Baah! I suppose it will not now be long before cables are old hat and radio or some other transmission is king.

I find it very useful to have extensive boards upon which you can set out basic information about your equipment for ready reference and one upon which you can scribble and clean off. There are all manner of ways to choose your target for the night, and there can be nothing more frustrating than to get home late with a wonderful sky and not know what you are going to shoot. Many people will go to their software, but as an extra aid I have a sky chart upon which I have posted mini pictures of objects of interest. This enables me to go straight to the hour angle for the ideal ascending target of the right size for the current set-up of the telescope. I also find it useful to have a list of targets on the wall, constellation by constellation. Do not forget to list the size of the object so that you know whether it will be suitable for the format in which you have the telescope that evening.

Neighbor Relations

Figure 5.37 shows the arrival of the mature sycamore tree. This was duly planted and left unguarded despite my remonstrations about the horse protection fence they were contracted to erect. "We will do that after the weekend." By Monday the horses had had a lovely feed on the bark so they had to return with a new tree and start all

Figure 5.37. The sycamore tree arrives.

over again. The adjoining neighbor to the one for whom the tree was planted rang up when he saw the tree appear. He is a high level executive of a cable TV company and had been supportive of my observatory project. "What a shame about the tree! I very much enjoyed the view of something which promised science." The good news was that I was able to persuade one of the complaining neighbors to water the trees during the sensitive first few months.

CHAPTER SIX

Telescopic Equipment Update

Progression

I had progressed from a 3″ Zeiss Jenna refractor to an 8″ Celestron reflector to a 16″ Schmidt Cassegrain LX200 reflector, and I also had a small Meade portable ETX. The 16″ Meade is a fine robust instrument, and in moving from the slide-off roof shed to the dome I took the opportunity to change from an altazimuth pier to an equatorial one. This provided the ability to piggyback another scope, and I chose a Takahashi FS 128 5″ refractor with a focal ratio of 8, which would enable a considerably wider field of view than the Meade at prime focus.

The Takahashi was quite a weighty instrument with its counterbalances, and I was somewhat concerned about the effect it would have on the Meade. However, there was no need to worry. The Meade was very happy with the extra load, and I felt that the balance of the instrument was even improved by the addition. Physically aligning the two scopes proved difficult, since I had opted for the most robust method of attachment and could only amend the pointing with shims. In looking for a very rigid connection between the two telescopes I had been hoping to guide with the Takahashi when imaging with the Meade through a hydrogen alpha (HA) filter, since this filter diminishes the amount of light received by such a large amount. My experience was that guiding the tracking in this way was never as good as when guiding with the Meade. I have no reason to think there was any deviation between the scopes, so that maybe the power of magnification of the Meade was the important ingredient. On the other hand, if you reversed roles and guided the Takahashi with the Meade, then you had superb tracking because of the superior magnification of the tracker.

The 16″ Meade is a good, solid tool that did me proud service for quite some years, although the electronics needed looking at from time to time. So robust was it that it had handles that you could actually hang on to. The corrector plate kept out the zombies—I was always nervous about cleaning this for fear of doing more harm than good. In the years that I owned the instrument the plate was only cleaned once, professionally, apart from the blowing off of dust and debris. There could be a little "flop" by the main mirror, but in time you got to know how to handle this. Over the years the telescope gave me very good service, and I managed some quite acceptable images with it when things went right. The telescope had a "wow" factor from its appearance (Fig. 6.1).

Having to play with the electronics board made me nervous (Fig. 6.2).

In the fullness of time the LX200 did have a niggling problem. It developed a very minor "throb" that occurred at about 2-minute intervals, which is about halfway through the periodic error of the machine as it hunts back and forth. This is nothing to worry about for imaging solar system objects but a total disaster for deep space. For months this persisted despite much help from the dealers in London. I made extensive logs of the malfunction but was unable to get to grips with it and was in touch with Meade's Technical Department but no one had a solution. I was about at the tearing my hair out stage when a new face appeared on the Meade scene. He was a character who had just sold the business he had created building wine cellars for the rich and famous in Beverly Hills and he was my night in shining armour. He would replace all the electronics and all the gears free of charge. Since the telescope was several years out of warranty I leaped on the offer with sincere thanks for his bounty. Everything now worked as it should have.

Figure 6.1. The 16" LX200 with Takahashi FS 128.

Figure 6.2. The LX200 electronics board.

My telescope time is usually taken up with a camera attached, but I found it interesting to look at solar system objects through each of the two telescopes simultaneously. Lunar objects seemed very slightly better defined with the Takahashi, whilst the planets were marginally better with the Meade.

RCOS and Paramount ME

There was chatter in amateur astronomy circles about some new equipment that was giving exceptionally good results in deep work. When one of the modern Ritchey Cretien telescopes (of Hubble ST-type design) was coupled with a Software Bisque Paramount ME Robotic German Equatorial mount imagers were producing Deep Space pictures of exceptional quality. I wanted one! Having researched the makers of the fabled RC telescopes I decided to run with RCOS, who were quick to supply me with full details of their product range and pricing; a local astronomer Adrian Catterall knew the manufacturer, Brad Ehrhorn, well and encouraged me to proceed in this direction.

For the telescope there were all manner of options, and I decided to order a 16″ F9 instrument complete with Aries optics, made in the Ukraine, Losmandy Dovetail plates, and RCOS Precision Instrument Rotator, an Astrophysics ×75 telecompresser (for wider field of view), and an RCOS Telescope Command Centre. The telescope was to be supplied complete with heated secondary mirror with which robotic focus could be achieved. Also included were three fans to bring the machine quickly to ambient temperature. The TCC provided the facility to robotically control the temperature of both mirrors (the primary via a Kendrick heater), to focus, to apply power to the fans, and to activate the PIRotator.

Telescope Assembly

There is the inevitable delay in the mirrors coming through from the Ukraine but eventually DHL turned up with the boxes: you have to build most of the telescope yourself! Adrian Catterall knew about these things, and thank goodness he agreed to come over and help. First we displayed all the parts, then set about installing the motor for the secondary mirror, as shown in Figs. 6.3 and 6.4.

Figure 6.5 shows the primary mirror ready for insertion, and Fig. 6.6 is a section of the telescope.

I had downloaded the instructions a couple of weeks before, and they intimated that the two mirrors should be aligned in accordance with two marks. Search as we might we could not find these marks, which sounded fundamental to the operation, so came to the conclusion that we should remove the secondary mirror to ascertain where was the elusive mark. All went well until the last moment, when a shower of tiny ball bearings suddenly appeared. This was probably not a desirable event! There was no alignment mark on the secondary, and it took me a couple of days to learn that

Figure 6.3. RCOS parts.

Figure 6.4. The secondary mirror motor.

Figure 6.5. RCOS assembly.

Gordon Rogers RCOS 16 inch f/9 Ritchey-Chretien

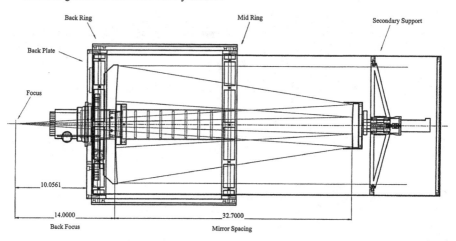

Figure 6.6. RCOS section.

since downloading the instructions the need for rotary symmetry had been dispensed with. Nevertheless we continued with assembly of the parts. I saw the primary mirror as a work of art. It was heavy and beautifully shaped and coated in Enhanced Aluminium. It was to be cushioned on the back plate slotted over the central hub and held in position by pads tightened against the glass by Allen keys operated from the central orifice. Once in position, it would be further secured by a retaining ring threaded onto the central hub. The primary baffle tube was wound onto the central hub.

I now had a telescope, but the secondary needed some sorting out. One thing I have learned is that if you take something apart, make sure you can put it back in the same way. Out came the paint pot and each section to be disjointed was given matching marks. Off to Flagstaff, and in no time the mirror assembly was back with its due ration of ballbearings.

Setting Up the Scope and Mount

Conveniently, the Paramount ME soon arrived. The bad thing about this mount was that it was of German Equatorial type, and thus the sky had to be viewed in either a westerly or easterly format. To change from one to the other entails turning the telescope through 180 degrees just after it has reached the best bit of sky, the meridian. The really good news is that, when expertly set up, the mount will track the sky so well that some imagers achieve stars that stay beautifully round during exposures lasting 5 minutes or longer. I had not quite got to this level, but I was pretty close. The mount has a combination of sophisticated engineering and a very simple routine for aligning it with the sky. Having released locking nuts it was a simple matter to adjust the azimuth and declination of the mount. I found it useful to stick markers on the adjustment knobs so that I could assess what was the effect a particular motion.

I had one serious difficulty to overcome. Since the LX200 was equatorially mounted the pier was constructed 2 feet 3 inches south of the center of the dome. A German Equatorial scope should be in the center of a dome. It was no simple thing to calculate the precise tolerance required, and I certainly wanted the telescope to be as high as possible because of trees. Having done my sums I went to my local steel fabricator, and he made me a 1″ steel extension plate to permit the mount to be centered.

Now I had all the kit, so I drummed up some fit young men and we put it all together. The most difficult part was sliding the telescope onto the mount, but with a bit of wriggling we finally had it in place. I was somewhat concerned at the rather massive leverage being applied by the addition of very heavy weights to a long counterweight shaft. There was no need to worry, as I now know the engineering to be so good that users plant all manner of scopes on a Paramount ME.

With the RCOS in place I was now facing a big word "Collimation." I had heard that Ritchey Cretiens were brutish to collimate and had acquired a Takahashi collimating scope, which was recommended for the process. With Adrian's help, we achieved a good collimation, and I find it hard to believe that collimation is still good after three years. In that time, the secondary has been to Flagstaff and the primary mirror has been in and out several times.

An interesting feature of the Paramount ME is that cables can be run through the mount so that they will not catch anything when the scope is being maneuvered (Fig. 6.7).

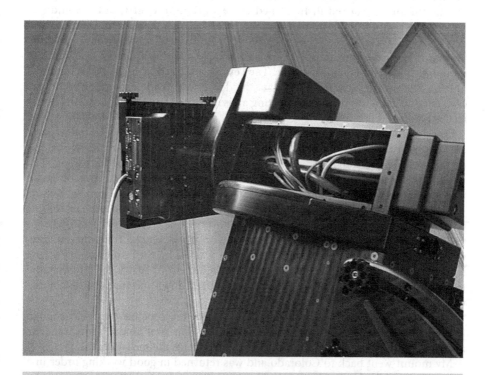

Figure 6.7. Paramount wiring.

There were connecting points for power and other utilities to be fed through the mount. My experience is that systems perform better if straight through cables are employed. Every extra cable junction seems to make the operation a little less predictable. The mount has a hand controller that enables movement in either plane to a progressively increasing power.

Problems!

There were some flaws! The dome motor assembly extended some 12″ from the face of the dome, and I should have had the mount half an inch lower. Not long after installation, on a fine summer's afternoon, I decided to become acquainted with the new equipment, first aligning the finder telescope before switching on any electronics. To do this I focused on the Didcot power station, some 18 miles distant, where there are some distinctive towers, pylons, and chimneys. The seeing was good and the clear magnification amazing. Next I thought I would do some experiments with the mount, having firmly in my mind the "no-go" vector of the dome where the lifting motor was located. Yes, you know what is coming next! There was a party in progress at the house next door and much female laughter. A dome acts as a huge sound trap and amplifies the volume. I was distracted and then there was a crunch. Idiot! I had driven the telescope into the motor. Quickly switching off I disengaged the mount and re-started it. It seemed to be working properly, but I would not know for sure until that evening. At night, when homing the mount, it was several degrees out. It was not easy to write to Software Bisque and confess what I had done to their magnificent machine, but they were very sanguine about it. Just send the mount back and we will look it over! They even tried to console me by saying I was not the first one to have tried to rotate the dome with the telescope. Steve Bisque even tried to calm me further by reciting a disaster they had had at Golden. Here is an extract from his e-mail.

It was a calm evening in the mountains just west of Denver, Colorado, and we were doing some Paramount testing. We have two buildings at our observatory; one has two roll-off roofs split in the middle and one rolls off to the east, one to the west. The other building is a converted pre-assembled 2.5 × 4 meter shed that runs on metal tracks about 6 meters long.

A storm appeared almost immediately from the north and found the east roll-off portion fully open and tilted just right so that it became a sail. As the 150-kg roof took wing, it headed straight for the building on tracks. At the time, the building was tethered towards the north and fully closed, waiting for the cement to dry on the large pier poured to hold our 20″ telescope. When the roof struck the building, the tethers broke and the building quickly gained speed down the tracks, aided by the wind gust.

"The closed doors on the mobile building struck the concrete pier with enough ferocity to explode both of them into many pieces. The roof landed some 3 meters away and dug a 0.3 meter crater in the hard soil. Luckily, no one was hurt."

My mount went back to Colorado and was returned in good working order in a few weeks. Luckily the damage had not been too serious.

Another difficulty was that with the combined weight of the mount and telescope there was a spring in the supporting plate: not good for deep space! Robert Dalby of Astro Engineering came to the rescue and manufactured a splendid cantilever unit, which provided a totally robust platform for the telescope centrally in the dome (Fig. 6.8). Figure 6.9 shows the Paramount ME face on with the readily accessible azimuth and declination adjustment mechanisms.

In the fullness of time the telescope was at last set up on a firm central base in a position where it was immune from interference from the motor casing (Figs. 6.10 and 6.13). The first thing to do was to align the mount with celestial north. There were various ways of doing this, but I opted for the drift method using stars on the meridian and in the low west since east was not available due to trees. Gradually the star drift diminished as I made the very simple adjustments to the altazimuth and declination knobs. Early on I learned to make a big adjustment in each plane and to write down the effect it had so I was certain of the direction of the star movement that each action would cause. I was doing this visually with an illuminated reticle eyepiece and gradually increasing the duration of the sightings. (If doing this today it would be less tedious to do it through the CCD camera.) Eventually I was happy that I had the best alignment I could get.

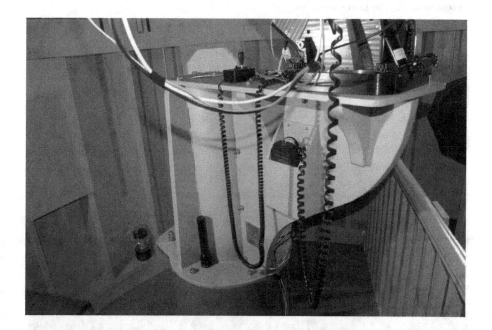

Figure 6.8. The cantilevered offset.

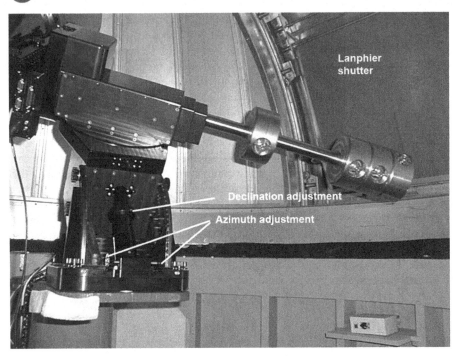

Lanphier
shutter

Declination adjustment

Azimuth adjustment

Figure 6.9. The paramount ME.

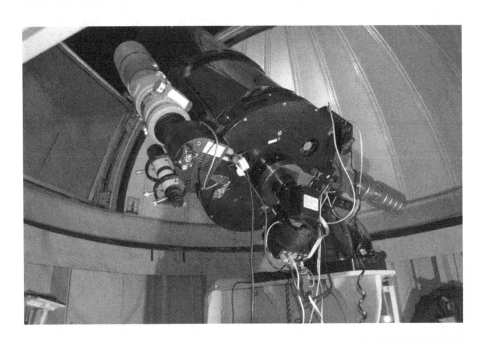

Figure 6.10. The RCOS with FSQ.

Setting Up and T-Point

When turning on the Paramount it had to be "homed" to a known spot in the sky a little to the west of the meridian, which it finds automatically when so commanded. You get a beep when properly settled and ready to go. When finished for the night, the scope can be "parked" in any desired position, but it is practical to park in the vicinity of the homing spot. The mount gives off a fair amount of electronic noise, but I find this fully acceptable because it tells you the machine is performing its function.

Stage two in the setting up process was to instigate the T-Point software procedure. Even the best of telescopic mounts have mechanical errors, and the alignment with celestial north will not be perfect for the whole rotation of the scope. There will be errors in the perpendicularity between the Right Ascension and Declination axes and various other gremlins to make the pointing less than perfect. This very clever and easy to use program provides a way to map the sky as it is seen by your telescope system. I used *The Sky* software (more later) to control the telescope, and the first step was to link T-point to *The Sky*. *The Sky* gives you a choice of luminosity of star to be viewed, and my first step was to limit this to a magnitude of 6 to avoid selecting a star close to the chosen one by mistake. Then I manually directed the scope to a spread of stars across the whole observable sky. The more you can include the better the result. With the scope pointing at an identified star you click the "map" button in the box, and the software records (a) The coordinates of the star you mapped and (b) the coordinates of the point at which the scope was aimed. This information can be accessed as you proceed, and you can watch error trends develop in various parts of the sky. The T-Point software provides many diagrammatic ways of viewing the results; for me the bulls-eye showing the deviation of each mapped point is the one I refer to most. Having completed your sky mapping survey the software gives a large array of terms that can be applied to arrive at the T-Point model for your telescope. I must say that I cannot begin to understand many of the more subtle terms, although I tried quite a few before coming to a conclusion. Just seven of the more basic terms for alignment errors and fork flexure gave me the best model I could achieve. With this applied the mapped points nearly all fell within a 1-minute central ring in the "bulls-eye" compared with around 17 minutes at the outset—quite a result! (Figs. 6.11 and 6.12).

Figure 6.12 also shows the terms applied. I carried out several T-Point runs and remember the frustration when the mount decided to "freeze" when changing from west to east, thus loosing an hour of mapping.

Piggyback Telescope

When I came to mount the Takahashi FS128 on the RCOS, I found the balance point required that it be located in a position which entailed an inconvenient protrusion behind the RCOS. The field of view with the FS 128 was only about 30 minutes, and I had heard very good reports of the Takahashi FSQ a small refractor of 106 mm aperture, 530 mm focal length, and an *F* ratio of 5. The

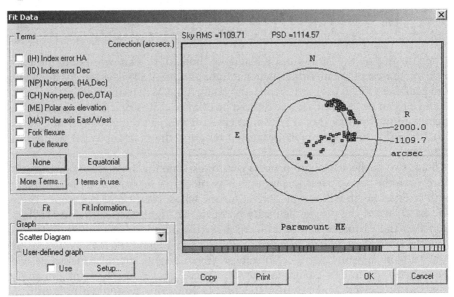

Figure 6.11. Initial T-point results.

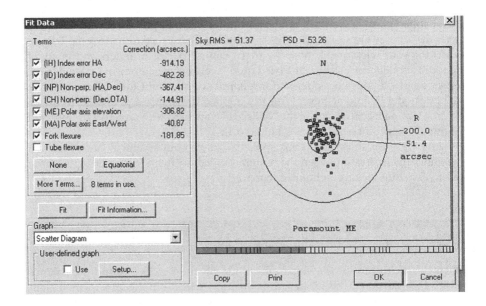

Figure 6.12. T Point results after application of seven terms.

comparisons between the fields of view of the RCOS and the FSQ when the SBIG ST10ME is installed are

RCOS prime focus	$14' \times 9'$	dimension per pixel 0.38 minute
RCOS at F6.75	$18' \times 12'$	dimension per pixel 0.51 minute
FSQ	$96' \times 65'$	dimension per pixel 2.65 minutes

I cannot sing the praise of the FSQ too highly. It has amazing light grasp qualities and the optics are of the finest quality. Image the Milky Way through a Hydrogen Alpha filter and you will be delighted and surprised with the amount of detail defined in just 10 minutes or so. I am so pleased with the FSQ that I tend to set up to image with that when the Milky Way is around and through the RCOS when it is galaxy time.

Adaptive Optics

I have an SBIG AO7 adaptive optics unit and consider it to be an essential device for imaging deep space.

The mirror of the instrument is introduced into the light path at an angle of 45 degrees and is controlled by actuators in a similar way to a loudspeaker. These actuators tip and tilt the mirror up to 40 times a second to fine tune the guiding of the light path to take account of residual guiding errors of the mount and even perhaps to mitigate in some way the effects of bad seeing. Compare images with and without

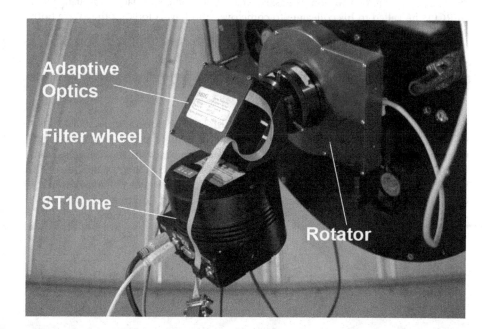

Figure 6.13. AO7 Adaptive optics unit and ST10ME with PIR rotator.

the AO7 activated and you will have no doubt which is which: the AO7 clearly gives a greater accuracy.

One little dodge I found was that the Paramount likes to pull rather than push. The way I am set up the mount pulls in the western sphere and pushes in the eastern; consequently, in the eastern cycle, I hang an extra small weight on the counterbalance shaft to redress the balance.

Mounting a Lens Directly to the Camera

Some imagers were getting splendid very wide-angle results by mating high quality lenses directly to the camera with the aid of a muscle plate by Steve Mandel. I managed to find a secondhand Nikon 400 mm lens, which gave conventional daytime images of magnificent quality at a good range. Piggybacked to the telescope the tracking was perfect because of the low magnification, although focusing seemed temperamental. I achieved some fair images but got nothing to shout home about and kept the lens for non-astronomical work. Returning from Costa Rica where I had taken photographs of everything from boa constrictors to sloths, the security people became highly suspicious about something large and long that showed up in the X-ray of my suitcase. Well done, I said; they missed that coming out at Heathrow. It is a camera lens. "Well, wa'do'yu'know" said the officer on examining it. Explaining that I had bought it for astronomy he asked what sort of telescope I had. On hearing I had an RCOS he yelled to his colleague to come over and have a look, as they were both keen astronomers. No problem.

Takahashi Binoculars

For visual delight I acquired a pair of Takahashi 22 × 60 fluorite binoculars with a field of view of just 2.1 degrees. They are actually a pair of small telescopes bolted together and give an awesome and colorful view of, for example, the Pleiades. From the bridge of a ship, at night, I invited the officer of the watch to view the fine detail that was visible, with the binoculars on a stand, of a distant vessel. Having viewed the sight he then looked at his radar screen and exclaimed, "My God that ship is 17 miles away!" The binoculars provide a marvelous means of observing comets.

CHAPTER SEVEN

The Opening Ceremony and "Sky at Night"

So now I had an observatory at the edge of Long Crendon village at co-ordinates 51 degrees 47 North, 01 degree 1 minute West. The name I arrived at was the Crendon Observatory—not very inspirational, but the best I could think of. Bob Antol (Chapter 14) named his observatory after his cat Grimaldi. I once had a tiny dog called "Whopper," so I suppose that could have been an option.

Sir Patrick Moore did me the honor of agreeing to open the observatory. This gave me the opportunity to return some hospitality to the professionals from Hertfordshire and London who had advised me, to the neighbors for their support, to the members of my local astronomy society and some prominent well-known amateur astronomers for their help, to the builders who put it together, and of course to the planning officers who had given me consent. Now, builders universally like a drink, and it was a damn close thing, but none of them quite fell into the indoor pond.

Kim Wilde, a pop star, came along. She is keen on space and has now branched out into gardening. On this subject she regularly appears on television, and she won a Gold Medal award for her Chelsea Flower Show garden. She found herself in conversation with Terry, the jack-of-all-trades builder. Making conversation she intimated to him that she was looking for a plant that would hold its color all summer. Terry's response "How about something in plastic?"

Sir Patrick's address was, as always, witty, and enlightened. In appreciation, knowing that he was having problems with his 1908 Woodstock typewriter, I gave him the oldest machine I could find, bequeathed to me by Margaret's aunt, a portable of pre-war vintage, but it did have a colored ribbon. Those of you who may have corresponded with Patrick will know that he does like the colored option although, with the Woodstock, the color with his favored machine does tend to "merge"; the "r" and the "v" usually do not function, and it is probably half a century since the typeface was cleaned.

Patrick seemed to like the observatory, and Fig.7.1 shows him on the balcony; a little later he telephoned to say that he would like to use it as a subject for one of his

Figure 7.1. Sir Patrick Moore with the author.

renowned *Sky at Night* programs. These programs were usually shot in the studio, but he would come out to me. This seldom happened, the last time being for the washed out eclipse of the Sun in Cornwall the year before. (No one who saw the eclipse coverage will forget the image of Patrick reporting from beneath his umbrella. He is that sort of guy!).

The program would take a day to make, and down came Patrick with his team of producer Ian, assistant Laura, cameraman Vince, and sound recorder Doug. I am no broadcaster, but Patrick is such an amiable character that it just becomes a conversation with him. There was a crude brief about subjects and times and I did my 2 minutes on one of them. "Very nice" said Ian, "but that was 7 minutes." I like to think I am not verbose, but this is just a measure of the way Patrick's smile encourages you to talk. Ian then instructed a change of scene and I should talk facing the wall. It did not take Ian long to call a halt and instruct me to address Patrick again. We paid a visit to the elevated balcony with its low door and, despite a caution; Patrick cut his head on his return. Margaret was quick to patch him up with a bandaid. This was before filming had finished ahead of luncheon. After the meal Laura was quick to rip the bandaid from his wound for the sake of "continuity." I was amazed to read in the Radio Times that the billing for the program said that Patrick Moore would "visit the Crendon Observatory to learn how deep-sky images of distant galaxies and nebulae are produced!" The great man is supposed to be learning from me? Patrick said he was pleased with the program, but I found it disconcerting to hear all the "errs" and "ums" that I introduced.

Sir Patrick's monthly program for the BBC has been in production for 50 years and is the longest running TV program worldwide by the same host. (Patrick has only missed one program, when he was laid low by a rogue goose egg). On April 1, 2007, Patrick presented the Golden Jubilee Sky at Night program in, typically for him, a very novel fashion. He played himself now, Jon Culshaw the impersonator played a magnificent Patrick in 1957, and Brian May played Brian May on Mars in 2050—hilarious!

I cannot let pass mention of Sir Patrick Moore without a few words about this amazing gentleman for those readers who do not know about him. He has a razor sharp brain, a rapid wit (asking him the other day if he had had any "out of this world" experiences he immediately retorted that he had been to Bradford!), and a presence that puts you at ease. He is a compassionate man who never thinks of his personal gain but always the welfare of others. He has no written contract with the BBC, just a gentleman's agreement. He could make a fortune by allowing himself to be featured in television advertisements, but he chooses not to do this, although he does appear for charitable purposes sometimes. He is the author of over 100 books in the astronomy and space fields and has done more than any other person in the UK to promote interest in astronomical activity for many decades. (I think he did stray into the science fiction world but that seems to remain behind closed doors.) Following the Boxing Day tsunami he co-authored, with Sir Arthur C Clarke, the book *Asteroid*; all proceeds of this were donated to the Sri Lanka tsunami relief fund. He has just written *Bang*, a complete history of the Universe. The co-authors of this book are the astrophysicist and renowned guitarist Brian May (Brian was the guitarist on the roof of Buckingham Palace at the time of the Queens Golden Jubilee celebrations) and Chris Lintot the astrophysicist and co-presenter of "The Sky at Night."

Chris Lintot did a show from the European Space Agency's Paranal Observatory. A Chilean amateur observer Daniel Verschatse took his small refractor to the site where-upon a queue of professionals appeared, all anxious to have a real-time look at the sky rather than through an 8½-meter electronic telescope with the latest adaptive optics.

Patrick could provide you with a Who's Who of the space world. He has met Orville Wright and Neil Armstrong on many occasions, including featuring Armstrong on a *Sky at Night* program. This was quite some achievement, since Armstrong rather shuns publicity. Patrick says it is such a shame that these two pioneers never met one another. He knew Wernher Von Braun well and questioned him about the slave labor at Peenemunde. Patrick is inclined to accept Von Braun's assertion that he did not know but should have known. Patrick is particularly close to Buzz Aldrin, who came over specially in 2001 to present Patrick with his BAFTA for communicating science and astronomy to the public. He has met Yuri Gagarin and Alex Leonov, the first man to undertake a space walk. He has also met me!

In the war, Patrick lied about his age to get into the Royal Air Force at just seventeen years of age. He was sent to Canada for a course in navigational training. While there, he managed to attend a scientific conference in New York. Einstein was there and, as an inveterate fiddler, wanted to know if anyone in the room could play the piano. Up steps Patrick, and Einstein asks him if he can play "The Cygnet." Of course! Now wouldn't a tape of that performance be something. I asked Patrick if he had had a talk with Einstein following the performance. Yes, a long one. What did you ask? "Could he define infinity?" He replied that he had never been able to do so in a way that would really satisfy him.

Patrick does not like to talk about his wartime experiences, but I hope he will not mind me saying that he was a Pathfinder Navigator and carried out a heroic deed while badly wounded.

Patrick is a man not to be messed with. Leaving the BBC one night, he was attacked by two thugs; they were left licking their wounds as Patrick went on his way! He could not report the incident to the police for fear that they might arrest him for injuring his attackers.

Patrick has enormous musical talent, and some of you may have seen his xylophone recital at the Royal Variety Performance before the Queen. He can write the most rousing march and a splendid Viennese waltz. One of my favorites, that was written and performed by him, is The Penguin Parade. Entirely gratis, he wrote the March of the Parachute Regiment. In pantomime he makes a wonderful ogre. Patrick has shaken hands with Rachmaninov.

For many years, Patrick was a stalwart of Selsey Cricket Club, bowling devilish spin with, on Patrick's admission, an ungainly and cumbersome run-up. Maybe the run-up paid a part in foxing the batsman, but he was very effective at taking wickets.

The extent of Patrick's fame was brought home to me when he asked if I had an image of a galactic cluster. I do have some, but I knew an American astronomer, Stan Moore, who had some that were far better than mine. E-mailing for consent to use them in *The Sky at Night*, Stan replied affirmatively saying that, as a boy, the first astronomy book he had read was by Patrick, and it would be a privilege to have him air his picture.

Just recently Patrick published his *Stars of Destiny*, a scientific and light hearted look at astrology. His conclusion was that it was all a bit of fun, but of course with no foundation whatsoever. Yes, you could make out that Sagittarius had the shape of a teapot but, equally, you could also call it a gas pump.

CHAPTER EIGHT

Cameras, Computers, and Software

Cameras

Now here is a field where the opportunities are endless. I have just touched the surface with the cameras I have used. However, I will restrict what I say to the models I have personally handled.

Camera Types

My first attempts were with conventional cameras, where you crossed your fingers and hoped you had the focus and exposure right, only to find out a couple of days later that you did not. There was all that business about being so careful with the shutter release mechanism and a discussion about which type of film to use. This all reminds me not to moan too much about a slight glitch with a CCD camera.

The digital cameras that I have owned are

Starlight Xpress 1CX027AL with an array of 500 × 290 pixels each of 12.7 × 16.6 microns.

SBIG ST6 T1 T2–241 detector with an array of 375 × 241 pixels of a massive 23 × 27 microns.

SBIG ST7 with a Kodak KAF-0400 detector having an array of 765 × 510 pixels each of nine microns: this camera incorporated the ST4 guiding chip comprising a T1 TC-211 detector with an array of 192 × 164 pixels of 13.75 ×16 microns. I recall how tiny was this chip for finding a guide star. I never imaged with the ST4 but the chaps who managed to have to be heroes!

SBIG ST8ME of 1530 × 1020 pixels. This camera had the Kodak 1603HE detector with pixels of nine microns. I upgraded this camera from the ST4 guider to the TC 237 detector of 657 × 495 pixels of 7.4 microns.

SBIG ST10ME has the Kodak 3200ME detector of 2184 × 1472 pixels of 6.8 microns and the guider is the TC 237 detector of 657 × 495 pixels of 7.4 microns.

The Santa Barbara Instrument Group cameras come with a desiccant cylinder to absorb moisture from within the camera. Every few months it will be necessary to bake the plugs to recharge the desiccant. It does not take long to learn the telltale signs of icing in the camera; this shows up in the shape of a peripheral deposit to one side or the other. You should be careful not to bake the rubber washer nor exceed the specified temperature.

I get rather squeamish when I have to take electronic things apart, and, of course, there is no one else to do it for you. Early on in the game you realize that it is an activity for loners. Much of the fine tuning of your equipment can only be accomplished when there is a sky and you cannot prearrange a helping hand. The chip in my camera had to be changed and Fig. 8.1 was taken during the process. Exposed are the color filter wheel and its motor, and the two triangular components comprise the adaptive optics unit. Positioning this unit fully square on the camera I have found to be a delicate operation and have learned to insert wedges of identical size as an aid.

Figure 8.1. Nervous times taking things apart.

There are many other excellent cameras, but I just speak of the ones I have used. There is a great variety of camera control options that are available, but again, I describe just the ones I use. All deep space imagers build their pictures by adding a series of individual exposures. There are many reasons for this, including accumulation of light pollution, interference by airplane (you will see the light strobe points) or satellite, by loss of guide star, by wind buffeting, by technical glitch, by cloud to name just a few. With multiple exposures at least you will have in the bag the ones that are not flawed.

Adaptive Optics

The AO7 adaptive optics instrument is available through SBIG, and I think it to be essential for deep-space imaging. It is not a camera, but to my way of thinking the camera is incomplete without it. The tip tilt mirror will activate up to forty times a second with a bright guide star and can hold on to the star within a 25/75% movement of the mirror. I think the best guiding comes when movement stays in the 40/60% band but cannot explain why the error can vary so much. Even with the best guide stars there is no certainty of low error. It can be quite unsettling to watch the error creep toward 25 and then recover.

Stack and Accumulate

Before the advent of the guider chip, I used the stack and accumulate provision in the SBIG software and acquired some very satisfactory images. The method here is to set the duration of the exposure for the length of time that you think your mount will hold station in the sky and then ask the software to align and add the images. This is still a good method of achieving very satisfactory pictures with a somewhat unreliable mount.

Computers

As technology improves I have gradually upgraded my machines and now have two set up with Windows 2000. Windows XP might be a better program, especially in the capacity that it enables you to automate tasks, but I am very stubbornly ensconced in the "If it is running OK do not try to fix it" routine. My main computer was professionally built for me, runs at a moderate speed, and is used when I am working up an image. The secondary computer I use to download the image from whichever telescope is employed. The computers are networked, and the secondary machine downloads directly to a folder in the first. If guiding the FSQ with the RCOS the main computer handles the guiding. The use of two machines makes life so much easier. Running *The Sky* on the main machine I can watch the effect of scope pointing adjustments as they happen. Also, having taken a focus image via the second computer, I can put the image under the microscope in the main machine as the next focus image is being delivered. I have the power to assemble mosaics of up to around six images but I doubt the machine would handle much more than that. For his glorious Andromeda galaxy image I think Rob Gendler combined around 16 images! Of course chips are getting ever larger and computers so much faster: which makes me think that amateurs might soon put together an all-sky survey (might be tough for the guy who lands Hydra). A laptop is handy when you are tinkering in the dome by letting you observe the immediate effect of commands such as when you have a malfunctioning filter wheel and want to see exactly what is happening. I also use the laptop when carrying out my T-Point program.

The Sky Planetarium Software Program

I find this an awesome program for keeping in touch with the night sky. Having uploaded the software the first step is to dial in details of your location and time. Personally, I stay on UT throughout the year; maybe part of the reason for this is that I object to politicians telling me at what time of day I should get up. Summertime also robs me of 1 hour of imaging sky.

You set the parameters of how you want the screen to appear and what should be displayed. You can actually map your horizon so that you see exactly what sky is available to you. You have choices about displaying the ecliptic, the Milky Way, equatorial and horizontal grids, and the direction from which you look at the sky, my preference being the zenith. You can toggle labels on and off to determine which objects are shown and which are hidden. At the touch of a button you can hide or show stars, double stars, variable stars, galaxies, star clusters, or nebulae. You can feed in the parameters of all the scopes you are using and call up the field of view of one, or more than one, that you wish to have displayed. If the camera is a dual-chip camera, the secondary chip field of view will also be shown; this is extremely useful in helping to find that elusive guide point when imaging in an area of sky with a paucity of stars. Of course, professionals "cheat" by firing lasers into the upper atmosphere and creating an artificial star by bouncing back light off sodium crystals. London Airport authorities might not approve if I did that.

Call up the program, and there is the night sky displayed before you just as it is at that moment.

Figure 8.2 shows the aiming point of the telescope with the "steering wheel," the target with a "bulls-eye" and segments approaching the meridian in differing shades). If you have not pre-selected a target, just draw a box around your ideal piece of sky for that night, click within the box, and it is magnified to the size that you have embraced. If you see something interesting there, click on it and up will pop a box telling you all about it. Included will be the coordinates and, if a star, the magnitude, the spectral class, and quite likely a spectrum and distance. (A refinement (?) in the latest program is that distances are often shown in AU, just with so many more zeroes than the previous scale of light years).

A nice touch is that points of the compass will be shown on the chart. I have always found it hard to keep my bearings in the sky. There is a "find" facility so that by typing details in the dialog box of the object you wish to locate you can have its position

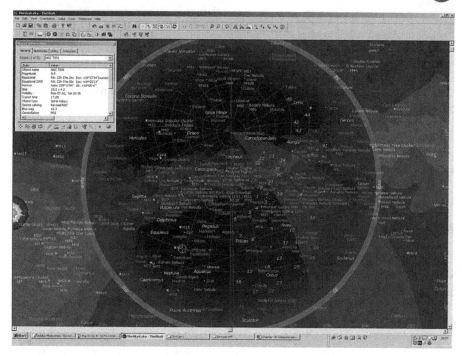

Figure 8.2. "The Sky" in zenith mode.

displayed. For some of the lesser catalogs you will need to learn the format in which the program needs the details to be expressed.

Prominent deep-space objects are often represented pictorially; this is a big help in deciding the orientation of your camera, and it is possible to integrate your own images with *The Sky*. There are tools for finding eclipses and conjunctions and an interesting time-skip mode whereby you can observe an accelerated solar system charging through the sky. There are a large number of telescope control software options but, not being an experimenter, I have always used *The Sky* and am very happy with it. There are motion controls to the four points of the compass at varying amplitudes. There are settings to control the speed at which the scope slews to its target. There is also an option to adjust the acceleration on commencement of move. Connect the scope with the software, home the mount, and you are ready to go. With a German equatorial mount *The Sky* draws a colored segment so that you are in no doubt about the approach of the dreaded meridian. (There are guys who have orchestrated the switch from East to West in robotic mode, but I am not in that class). I should say here that they tell me that should the mount reach its limit a few degrees past the meridian, the mount will carry out its rotation automatically, but I would rather control this since not all my cables run through the mount and something could snag. It is possible to type in specific celestial coordinates and command the telescope to go to them.

Target Finders

So you have a clear sky and want to get cracking, but at what target?

In my experience you cannot be too fastidious in preparing target lists including, as an essential component, the size of the object. You do not want to waste sky time changing the format of your equipment.

In addition to guides and charts displayed in the computer room there are a number of ways to gain more alternatives. If you are a member of a camera user group the images being displayed there will normally be currently available, and there are software and Web applications. Two that I use are:

NGCView

This program gives an easy-to-select range of definitions by which a target can be sorted, including position, brightness, size, shape, type, name, and observing history. Objects conforming to the defined specification can be called up and appropriate information displayed together with a chart showing the elevation from your site and the position in the constellation in which it appears.

Deep-Sky Browser

Located at: http//www.deepskybrowser.com/cgi-bin/dsdb/dsb.pl, a free program, constructed by an astronomer, provides an expansive search regime to look for objects by name, position, constellation, or type. It is also possible to search by catalog, and this base is indeed comprehensive. For example, in Cygnus, there are listed 2,460 pages each of 50 items! Thank goodness you can narrow this down by type. If specifying planetary nebulae the number drops to 12 items.

Periodic Error

Periodic Error Programmes

The 16" LX200 did not have the ability to train the periodic error of the scope, but the smaller instruments do. The Paramount ME has this facility, and a number of software applications are available to fine tune periodic error. In the right hands (not mine!), I understand it is possible to bring this error down to a very small number.

Maxim

A word here about preparation before discussing the functions for which I use *Maxim*: it is necessary to "clean" images before processing them. This entails subtracting (a) the "dark current" and (b) a "flat field," being an image of the defects in the optical train.

The colder a CCD camera can run, the less the dark current, so that I always aim to run the camera at the lowest temperature that can be consistently maintained. Professionals, and some amateurs, go to great lengths to cool their cameras in the search for perfection. It is possible to plumb in a water-cooling system to a SBIG camera to gain extra cooling, but the downside of doing this seemed greater than the upside. I will usually take around ten images with the camera closed and average them in "median" mode to arrive at my library dark masters. This process needs to be done for the whole range of images you are likely to acquire and, for me, includes images at 0 degrees, minus 10 and minus 20 for exposures of 60 seconds, 300 seconds, 600 seconds and 1,200 seconds. Quite a lot of exposures, but on a cold cloudy night you can retire with the camera programmed to do its stuff. For the flat fields you again need median-averaged exposures to the value of about two-thirds of the capacity of the camera. These can be taken at dawn or dusk. Personally, I have constructed panels that fit over the telescope: for daylight a polystyrene one that diffuses and subdues the light and for the hours of darkness an opaque polythene that picks up diffused light reflected from the dome. Although the dark fields are constant for all images the flat fields will, of course, vary if the configuration of the telescope is changed in any way. The purist will take new flat fields for the work of each night, thus compensating for the fresh work of any little insect.

After some degree of experimentation I decided to use *Maxim* for camera control: I must be quite an aficionado as, in order to run it on two machines, I had to purchase two copies. I also used it for initial calibration of images. Having acquired an image I use *Maxim* to subtract the dark current from my library of dark images together with my flat fields. At this stage I go through the "remove hot pixel" routine and carry out an automatic "Bloom removal" routine if there are bright stars in the image that have exceeded the well capacity (Figs. 8.3 and 8.4).

Having filed the "clean" images I then identify two stars in each exposure and call for them to be aligned. Having done this I ask for the images to be combined using the sigma reject plug-in to *Maxim* by Russell Croman. (There are a number of available software solutions for combining images; this happens to be the one I know and like.) I am usually building an image from red, green, blue, and luminance exposures. The combined luminance I will bring into CCD Sharp (more later) to endeavor to de-convolute it a little and then return to *Maxim* and see what effect the "Digital Development" routine has. There are all manner of settings that can be applied, but I find that I often settle for something in the range of FFT low-Pass, FFT Hardness–Medium and cut off 2.5%. This can have quite a dramatic effect and is akin to a multi-band Photoshop treatment, but I like to compare it with what I get in Photoshop. The other *Maxim* facility that I use regularly is the color combine one. In stating the intensity of colors I usually find I have to scale back the green. I do not whether this is an idiosyncrasy of my set up or a normal outcome.

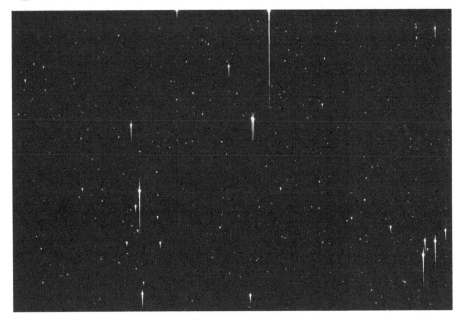

Figure 8.3. Image with severe blooming.

Figure 8.4. Bloom corrected in Maxim.

Sometimes *Maxim* gives me the better color combine result and sometimes Photoshop. I also try *Registar*, which is especially good at pinpoint alignment when images have differential rotation, having been taken over several nights. One thing I will say is that I find color composition is the most difficult aspect of trying to achieve a decent deep-space image.

Having connected to the camera the first step is to choose the temperature. This should be at the lowest level that can consistently be achieved, since with CCD cameras the lower the temperature the less camera noise is created. Next comes focus. Having downloaded an image in "focus mode" you draw a small box around the chosen star, preferably a fairly bright individual star, call up a high-resolution repeating image of that small area, and then invoke the routine whereby you can ask for a large display to show the co-ordinates, the photon count and the FWHM (full width at half maximum) of the profile of the star. Also displayed is the actual profile, or emission line, which gives a quick view of whether you are adjusting the focus to improve the focus or go the other way.

At the time of good sky conditions you can usually snap to good focus quickly. With a more mobile sky the feel is more spongy, and I find it useful to traverse from one extreme to the other, noting down the points where you clearly "lose it." Best available focus may well be midway between these points; at low magnification decent images can be obtained, but at high power the bad seeing will downgrade the image. With the FSQ I like to download a 10-second high resolution image and then inspect the stars with the zoom window. Once this test produces tight stars I take a 60-second high-resolution image as a final check. (I have learned to do this following the many 600-second first images where focus has been just out.) Because of the low magnification through this instrument I never bin the pixels.

Target Acquisition

Now to the target for the night: first find it and center it, and "synchronize" the telescope if it needs adjustment. Picture composition is so important, and time spent thinking about this is time well spent. Get a picture of the object from the Internet; you will be surprised at the huge array of astronomical images that are available. In fact, they say that the Net was first driven by astronomy and naughtiness.

Now you need your guide star in a convenient position for the picture in your mind. You need the brightest one you can find and, in the Milk Way, you should not experience too much difficulty. When imaging galaxies away from the Milky Way things can be very different. If you have called up the Field of View in *The Sky* for the telescope you are using you will be able to see the available stars and you may well have to increase the duration of the exposures to pick up sufficient signal for guiding. In a night of good seeing I have guided satisfactorily once every 8 seconds but am so much happier if guiding is ten times a second. For galaxies you often have to search for a bright enough star, and this is where the electronic rotator is such a boon. Of course, if you rotate the camera other than for 180 degrees your picture composition needs to be reviewed, unless using a separate telescope to guide. In my search for a guide star I activate in 20 degree segments. Having found that elusive star I always make a drawing showing the effect that "up" and "down" moves have on the chosen star so that you can maneuver it on the chip with certainty. (Quite frequently having centered my galaxy I have abandoned the attempt to image it for lack of a suitable guide star.)

Calibration

Now it is time to go through the calibration routine so that the software knows the alignment of the camera. The previous night's calibration will often suffice, but, if the orientation has changed, calibration is required.

With the LX200 I found the process rather "hit and miss" and often had to make multiple attempts at different exposures and different commands for duration of move. With the Paramount I am surprised if I do not get the calibration first time, and I have never had to change the parameters for duration of move. I find that simply increasing or reducing the exposure time invariably overcomes any difficulty. I make a note of the initial guide star co-ordinates and only accept the calibration if the final dimensions are within two or three pixels of the first. On initializing the guiding it can be tempting to go for the fastest possible rate, but I have learned to avoid this. I make allowance for the diminution of the signal due to the odd hazy cloud that often seems to lurk about. It seems to be accepted practice that guiding should be binned 3×3, and this is the method that I adopt. It certainly increases the number of stars that are viable for guiding. There is provision to adjust the aggressiveness of the guiding; having experimented considerably I seldom change this from 10×10.

Commencing Imaging

We have focus, the guide star is locked on, and the adaptive optics are functioning, so now let us start imaging our deep-space target. Invariably long exposures are required, which are best put together by assembling a series of similar exposures. If there are bright stars in the field the saturation that they generate will limit the time for which the camera can be left open. For example, when making a mosaic of the very bright Orion Nebula, working out from the center, my exposures were of, 2, 10, 60, and 300 seconds. Generally speaking I find 10 minutes to be a suitable time for red, green, blue, and clear filters. With a hydrogen alpha (HA) filter I normally expose for 20 minutes.

Sequence

Maxim has a very useful tool for specifying an exposure routine. With "sequence" you can individually command which filter shall be used, the duration and number of exposures, and the binning mode. The "settling" time for the camera can be specified. It is possible to have the images slightly "dithered" so that prominent small noise defects are averaged out when several images are combined.

Maxim Tools

There are a whole range of tools in *Maxim*, including telescope control (which I have not tried). In addition to the "zoom" window I invariably have the "stretch" window open. Under "Kernal filters" is a variety of algorithms that you might try but only after initial calibration. For image acquisition I use the Flexible Image Transfer System (FITS), the method widely used for astronomical images. The "FITS header" drop down gives exhaustive information about any FITS image being displayed, including temperature, length of exposure, binning, time and date, and camera and pixel size after binning. There is a whole gamut of tools, including specialist filters, deconvolution, flattening background, gradient removal, curves, astrometry, and a comprehensive range of color tools.

CCD Sharp

This is a dandy little deconvolution algorithm that I always try on my combined luminance images, and more often than not it improves them by tightening stars and nebulous shapes a little. The number of iterations can be dialed in, and I find that two iterations often give good results. With an image taken in poor conditions, the software will accentuate the noise, so do not go down this road.

CCDSoft

This is an excellent and popular system for camera and telescope control with many similar features to *Maxim*. I think there is no doubt that you should have on your computer a choice of operating systems, if only for the reason of being able to cross check. If there is a problem in *Maxim*, does it persist in *CCDSoft*? The first bit of detective work must be in trying to establish where the fault lies before you can attempt to remedy it.

Adobe Photoshop

I run Adobe *Photoshop CS* version 8. What a program! You could continue to learn about the nuances until the cows come home. There is such an enormous volume of tools to extract that extra miniscule piece of latent information from the image you have taken that the learning curve is endless. All I propose to do is to run through the basic routine that I adopt in processing an image. There is so very much more that the expert will do in fine tuning his or her image. If you want to get technical I recommend *Photoshop Astronomy* by R. Scott Ireland. This is an in-depth look at the application of *Photoshop* tools for astronomical imaging. Now back to mere mortals:

Since I acquire images in FITS format the first thing to do is to download the European Space Agency's free FITS Liberator plug in for Adobe. This will enable the stacked FITS images to be opened. I have also purchased Neat Noise Reduction plug in, which can often help to smooth an image taken in less than good conditions. The color image will frequently benefit from the application of Neat.

Another plug in that I have purchased and use regularly is Russell Croman's *Gradient Terminator*. Casts over filtered color images caused by the moon or light pollution can be quickly rectified in the simple procedure offered by this program. One thing remains clear; however, no matter how many rectification systems you have the best images are obtained on moonless nights of good seeing and transparency with minimal light pollution and excellent guiding.

Having opened the unfiltered CCD sharp processed image I first, in "levels," adjust the white and black points to bring out some of the detail. I am careful not to be too heavy handed and never attempt to stretch the image "all the way" in this configuration. Now to "curves," where the possibilities are endless. I like to make quite a number of small adjustments, gradually stretching the image until I like what I see. It is a good idea to go a step too far because the "history" pallet gives you the opportunity to step back to most previous levels. Trial and error is certainly the message here. You may find that brighter parts of the image become overexposed; in this case, pick up the "history brush tool," designate the appropriate brush size and apply it as appropriate. Sometimes this will result in a dark "blob." in this event go to "edit" and fade the selection until you have the right balance. "Brightness and contrast" is a tool that is frowned upon by some, but I find that modest adjustment here can help an image. If imaging in the Milky Way with huge swathes of dust and nebulosity it can help to adjust the density of black in "selective color." To do this, you will have to convert the image to color mode, select black, and advance the density—trial and error once more. For the nebulosity go to "shadows/highlights" and play around. There are many options here.

Can you improve the image further? I like to try a modest "unsharp mask" filter. The history pallet gives the opportunity to rapidly check whether this step has made an improvement; often you will have a degradation. While in "filters," if stars are a little bloated, you might try the "minimum" application and fade the result as appropriate. A last step might be to overlay the image with a duplicate of itself, call up the "high pass" filter at around four–eight pixels, blend in "overlay," and reduce opacity if required. Then, in "layer," go to "layer mask—hide all," and with the "paintbrush" tool reveal the extent of the image that benefits from the treatment.

To demonstrate the effect of various tools, Fig. 8.5 shows a wide field image centered on NGC 7822 in Cassiopeia, downloaded into *Photoshop* and adjusted in "levels." Figures 8.6–8.8 show further adjustments.

Now that the luminance image is complete the hard work starts. I usually assemble the filtered color images in both *Maxim* and *Photoshop* and work on the better result. In *Maxim* the function is available to incorporate the luminance for an LRGB, but I prefer to work with separate color and clear filter images. The color image may look a little "blocky." If so, I try a "Gaussian blur" at around one or two pixels radius. If combining in *Photoshop* you need to convert the images to 8 bit and use the "merge" option to create the color. I find it usually helps to boost the "saturation," since the combine will be at around 50% and color tends to get washed out. The combination can be done through "layers" in *Photoshop* but is probably easier to accomplish in *Registar*, especially if the images are a little rotated due to being taken on different

Figure 8.5. "Levels" adjustment.

Figure 8.6. "Curves" adjustment.

Figure 8.7. "Selective color—black" adjustment.

Figure 8.8. "Shadows" adjustment.

nights. Invariably the quality of the combined LRGB will be lower than the clear filter image. If combining in *Photoshop* there is a whole raft of "blending" options in "layers," where endless experimentation is possible. *Photoshop* provides a myriad of tools and systems to get the best from an image. I use just a few basic ones.

Is the color balance right? The human eye, well mine anyway, cannot detect color in deep space, but there are technical ways to assess color, and top imagers will spend much time honing color.

Search the Web for your target and see the high-class images that are available. Maybe light clouds happened to roll across each time you exposed the blue filter, requiring you to boost it a little. If the image suffers from a gradient this would be the time to apply "Gradient Terminator." Try some of the "auto adjust" buttons and see what you get; a click in History will instantly take you back a step if the process is retrograde. You will most likely need to go to "saturation" again for a further boost. Go to "Adjustments, match color—density," and perhaps a boost here will help. In "curves" experiment a little; also, if the target is a galaxy or other item with an isolated position, pick up the left-hand eyedropper and apply it in different parts of the background, observing the result. With Milky Way nebulae the color is often too pink. Try "Adjustments—replace color," click on the color you want to change, adjust the fuzziness to suit, and adjust the saturation of that color. "Selective color" gives the opportunity to manipulate the hue with which you are unhappy and "Shadows-Highlights" controls overall balance.

Narrow Band Filters

The Milky Way offers a wealth of targets rich with nebulosity and which are especially suitable for imaging through a narrow band HA filter because of the vast amount of excited hydrogen gas. Though slow, the filter does "cut through" and pick up intricate detail. Being a red source many imagers include a proportion of HA within the red image by blending them in "layers." (See Rob Gendler's manual in the section "Information Resources.")I find this can rather swamp the image in red, distorting star colors, but Rob describes a way of restoring the colors by selecting them in "color range" and bringing back the hue.

Imaging Solar System Objects

The technology for imaging solar system objects has leaped forward in the last couple of years. I have some decent images of the Moon, but the best amateur images today are obtained with modestly priced video cameras coupled with high-quality telescopes, which need not be too big. I have not tried this but gather that thousands of images are acquired and scanned and the best hundred or so combined. I am astounded by the animation prepared by Damian Peach from Barbados showing a whole rotation of Jupiter in full color (address in Information Resources). Each of the frames, and I guess there must be at least a hundred, is of superb quality. When you think how quickly the planet rotates there is just a narrow time slot to acquire the welter of red, green, and blue filtered information for each image.

CHAPTER NINE

Wrinkles Galore

People ask me if I play golf. My reply is that I own a bag of clubs and walk around a course once or twice a week. It is good exercise, but I used to think it the most frustrating activity imaginable. Astronomical imaging wins hands down! There is no doubt that if it can go wrong it will go wrong. On top of this, being domiciled in central England, I have probably chosen one of the least favorable sites for the activity in the world (Seattle excepted—I see that Mark-de-Reget of that town has now gone robotic to New Mexico). Perhaps I may recite just a few of the mishaps that have befallen me.

Whatever Can Go Wrong

First I take a good look at the weather forecasts and study the weather horizon. Things look OK. Fifteen miles to weather of me is the Didcot coal-fired power station. The huge cooling condensers throw out vast amounts of steam, promoting their own cloud formation. I have been inside one of these condensers, and, amazingly, the walls are only 4½-inches thick. These clouds will dissipate before arriving at Long Crendon. Switch everything on, get the cameras to temperature, move to the target and set about finding the guide star and calibrating; fine tune the focus and off you go. I could not attempt to count the number of times when half an hour or so into the project the guide star is lost and is not going to come back because of a deterioration in the weather.

It is bound to happen that the one day in the week when you are partying the sky will be crystal clear. On one occasion last year the moonless sky was riveting as we walked home. Maybe it was one o'clock and I had imbibed rather much, but I had to get some of that sky. I got everything going and the first 600-second image was brilliant. I let it run in sequence and fell asleep at the computer. I awoke at about 3.00 a.m. to a lost guide star. What was that noise? Rain was belting down. Dashing upstairs I found the shutter was channeling the deluge directly into the telescope, which was at an elevation of some 45 degrees and was open ended. I immediately threw the close shutter switch, but then I apparently did the worst thing by switching off the mount, camera, and telescope. I then manually lowered the telescope, and out gushed a gallon or two of nice fresh and sparkling rainwater. Part of the mirror was bathed in it, but the electronics were dry. Thank goodness the dome had sheltered the business end of the telescopes and the cameras. With some trepidation I switched everything on the following day, and all was fine. The mirror was unaffected.

Before the small hours I have often picked up airplane lights but am lucky to be away from the main flight paths. A much more frequent event is to pick up satellite trails.

Figure 9.1 shows airplane lights, a trio of U. S. Navy orbs crossing the Andromeda Galaxy, and an exceptionally bright satellite, perhaps the International Space Station. I got upset when satellites at 90 degrees actually take aim at my target galaxy. There have been times, when imaging in Sagittarius, I have forgotten about a particular ash tree. (This problem can be overcome by plotting your horizon in *The Sky* software.)

With the LX200 re-collimation was a regular requirement, and typically the problem would seem to arise on the nights of best seeing. I learned here to draw a plan showing the direction in which each of the three collimation screws would draw the star so that you could work out what was required.

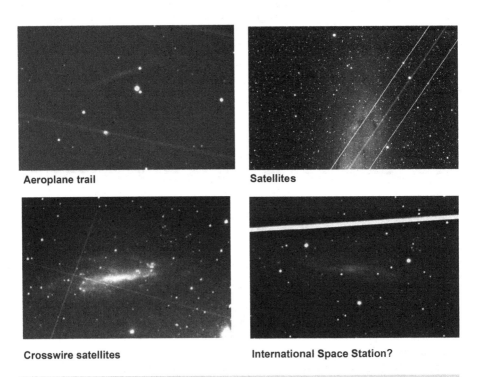

Aeroplane trail

Satellites

Crosswire satellites

International Space Station?

Figure 9.1. Satellite trails.

Amazingly, I have not had to re-collimate the RCOS, but I do have a problem with mirror flop when getting down to an elevation of 20 degrees or so. You can just watch the stars elongate as the telescope is depressed. I have had the main mirror in and out several times to try and overcome the problem. I have had the mirror so tight that I have deformed it but still no cure. The screws inside the inner orifice are awkward to get at, but I have adjusted these umpteen times through varying degrees of tightness with no luck. I just have to accept that the Trifid and things south of that are out of range. In any event, from 51 degrees North, exceptional seeing is required to get a decent image at that depression.

There have been times when the telescope keeps drifting across the sky, and no matter whether you restart, home, or synchronize the problem persists. I now know that the odds are that the software has decided to go back to home base and has moved you to Golden, California. I suppose this could lead to quite some difficulty if you live near Golden. With a problem like this always check the basics and make sure the computer clock has not thrown a wobbly.

I have been so busy working up an image I have forgotten how time has gone by and taken a wonderful picture of the inside of the dome. At other times I have left the observatory light on.

Earlier this year on a freezing night I lost power to the Telescope Command Center, which powers the heaters. Needless to say the sky is wonderful and on checking the fuses they are all right. An hour later I find that one of the fuse holders is faulty. It is at times like this when the computer does its "I am going to give you a blue and knock-out screen" and the only remedy is to kill the power and go through the interminable start up and checking display because you are the one who has misbehaved, not the wretched machine.

When there is a difficulty in communicating with cameras and accessories my experience is that the problem, in the vast majority of cases, is a connection. SBIG power and USB leads tend to inch out over time, especially if facing downwards. I now go through the routine of securing them up before I start. I have even been known to forget to switch the camera on! With this sort of problem change the USB lead, and if this does not work, try a different computer. If still out of luck, try different software.

Perversely guide stars often seem to arrive at the edge of the chip. If the image you are taking pretty well fills the exposure, it is worth trying to find a more central star to give the ability to adjust the precise pointing. I have had the experience of searching 360 degrees for a guide star and then finding a bright one. On calibration an equally bright companion appears, and you have no chance of reaching calibration as the software hunts between the two. Maybe the answer here is to go elsewhere to a star-rich area, calibrate, and return to one of the two available stars.

I had fun linking the ST10 to *Maxim* for the first time:
"What camera is it?"
"It is an ST10"
"No. Wrong!"
"Yes it is"
"No it is not!"
After 5 minutes of this I became fed up and lied: "it is an ST8."
"Well done, you have got it right at last!" The software then happily read the ST10's 2184×1472 array.

Having found and centered my target galaxy in focus mode I then went on to expose. The picture dumbfounded me. The field of view was 14′ × 9′ and the galaxy had shifted 4 minutes sideways. Back to focus mode and the galaxy was centered. Without moving the telescope, the camera was taking images of two different areas of sky. I tried everything I could think of to overcome this without success. Help at SBIG had never heard of the problem. Several days later there was a clear sky, and the camera worked as it should. Gremlins or maybe a poltergeist? More about this later.

Another peculiarity with *Maxim* was that it had a night (more than a year after the ST10 fiasco) of deciding that the camera was indeed an ST 8. By this time I had succeeded in persuading *Maxim* to accept the true type of camera, but the software persisted in producing images of 1530 × 1020 pixels. A couple of night's rest was the answer here, too.

The CFW8 filter has given me some grief from time-to- time. This has been for a number of different reasons. The first occasion was a sticking wheel, and the cause was a loose filter that I had not secured sufficiently tightly. The second time, sticking again, I eventually traced to the pressure on the drive shaft. This is adjustable, but the engagement needs to be exactly right. Not tight enough, and the drive will slip spasmodically; too tight and the wheel can bind. After a number of attempts I managed to get it right. I well remember one starry night: the wheel was working intermittently and one computer was out of order, so I went through the usual routines: check connections, use a different cable, and eventually took the wheel apart looking for a physical problem. Using the laptop in the observatory the wheel worked fine. At last, inspiration: I checked the parameters in *Maxim*, and the software had decided to change the wheel type from SBIG to Finger Lakes.

Bright stars close to the imaging subject can cause problems that are not readily seen. For example, the Horsehead has the magnitude 1.74 Alnitak nearby. Focus images may not show the intrusion of light from this star, but a 600-second image may well do so. In imaging this at prime focus I had to shift the nebula off center to avoid the impingement.

Recently I had set myself up for a long session on M78 with the FSQ. I had a really good guide star for the RCOS, and at around midnight I did the big rotate, made sure everything was OK, and retired with the scope to the extreme east of the shutter opening. This should have given me 2 hours of prime sky time. The result was one 10-minute image. There were a few straggly clouds, and one had knocked out the guide star. The software went mad searching for it. The images looked a bit like a child's sketch book.

I am not a solar observer, but I do have a crude cardboard assembly that enables an image of the Sun to be projected onto a card to see sunspots. On one occasion there was a very unusual linear marking—something I had not seen before. I had just e-mailed a friendly solar imager asking him to have a look when I examined the equipment and, for the first time, noticed the spider's web!

I have recited just a few of the frustrations I have undergone, and I do wonder if I get more than my share. Is there a deeper force at work?

Before moving to this house I had a neighbor, Peggy Nicholson, and she had previously lived here. Over the garden fence I asked why she had moved, since she was still in the village. She told me that it was a bit lonely at the house I was buying, but her words sounded rather defensive. Since at this time I was beginning to understand the workings of the female mind, I asked if she was hiding something. Well, if you must know, she said, there is a poltergeist. I asked what it did. She told me that it moved things around the house. Was it nasty? She assured me that it was not.

Peggy has another face. She is a renowned crime writer under the name Margaret Yorke. So here was the Chairman of the Crime Writers' Association of Great Britain warning me about a poltergeist. I took that with a pinch of salt. However, events, and there have been many, proved her to be right. Just a couple of examples: I was making a model in the utility room. Pausing to answer a call of nature, I returned to find the pliers I had just been working with had disappeared. "Margaret, have you moved them?" She says I am always mislaying things, but despite a determined search by both of us there are no pliers. Dinner is ready anyway. After dinner, returning to the utility room the pliers are in the center of the work station.

On another occasion Margaret dropped a diamond ring in a front bedroom. She could not find it. Neither could I, despite the most exhaustive search, including taking up the edges of the carpet. Six weeks later, leaving to go on holiday, sitting in the middle of the floor in a back bedroom was the said ring.

User Groups

I learned that there was a Yahoo user group for the Santa Barbara Instrument Group camera owners and thought it would be a good thing to join, to help me avoid some of my mistakes. Wow, about a hundred e-mails came tumbling through, so I thought: Let's get out of this! Before doing that, though, I thought I would just read one or two. There were e-mails from legendary guys, like Adam Block, Robert Gendler, Ron Wodaski, and Stan Moore and what is more they were giving constructive advice to lesser mortals, so it made good sense to maintain the barrage of e-mails. The title to the message immediately told you what was likely to be of interest to your set up. So, in the fullness of time, having taken what I thought was a brilliant image, I put it up for comment. "Do you think perhaps you should have processed it this way or that?" Yes, it is obvious when it is pointed out, and in fact the image was pretty dumb. There is so much to learn, and the reading of comments between members will open all manner of doors. I am always so impressed in the way that a newcomer will be taken in hand by the group and lead through even the basic essentials to get him or her on the right path.

The usefulness of the group was brought home to me in dramatic fashion. I had loaned my camera to Nik Szymanek and Robert Dalby, who had planned a week's observing on La Palma Island at the elevated Roque de Los Muchachos Observatory. Among others, the site houses the 4.2-meter William Herschel Telescope, the 2.5-meter Isaac Newton Telescope, and the Swedish Solar Telescope.

On the first night they telephoned to say that all was well. The following night there was panic: the camera had thrown a wobbly, and though both were very experienced with electronic cameras; they could not find the answer. A note to the group explaining that we had two up mountain imagers in difficulty and from Miami came a reassuring message from a member who knew the problem and sent an electronic patch to resolve it. All this happened within 3 hours of the difficulty being identified.

There have been some humorous notes. Tom Krajci from Tashkent, Uzbekistan, with the snow 12 inches deep all around his observatory, was worried about his body heat affecting the seeing. "I place a box fan so that it blows the warm air from my body away from the optical path. It makes a noticeable difference when high-altitude seeing is pretty good. Before the fan, my warm body would spoil the seeing, even if I bundled up well, wore a hat, held my breath etc." True dedication!

A German member put up an astounding image of M83 in Hydra. He had taken the image from Namibia. The black and white information had been acquired the year before and the color the following year. There were glowing accolades regarding the picture from all around the world. One top-level American imager admired it but voiced the view that perhaps the colors were a fraction strong. Now, this member had spent days, maybe weeks, nurturing this image and the comment got to him. Back came a reply refuting the criticism and demonstrating why he would not accept it. There were links to two sites. The first was to a German Christmas tree sparsely decorated with golden balls. The second was to an All-American Christmas tree decorated in grand fashion, with lights, baubles, chocolates, presents, and all manner of bric a brac to such extent that the tree had all but disappeared. These two guys stay good friends.

Eddie Trimarchi from Australia put up a black and white image of an intriguing nebula, NGC1929, of quite weird appearance. I asked if he was planning a color shot, and before he could reply Stan Moore (for whom I have the utmost respect as a higher intellect being) implied this would be a waste of time. At first Eddie succumbed to the veto but several of us persuaded him to do this, and it was certainly rewarding to see the result. Stan is a purist, and I think his point, which I do not dispute, was that the resolution of detail was better recorded in black and white without the addition of color.

Adam Block was running an observer instruction program for National Optical Astronomy Observatories from Kitt Peak, Arizona, at an elevation of 6,875 feet with very stable seeing and low humidity. Under instruction from him, his students produced a dazzling series of images on a frequent basis. He put up a good image of the galaxy NGC 2336 in Camelopardalis, and intimated that this was the furthest north that he had ever tried to image. For me, the galaxy was overhead, so could I get something that might compare with the Block image? Even with similar equipment, the same exposures (although his camera was 10 degrees colder), and a night of good seeing, this was not to be. The luminance was not too far adrift, but his color signal was much stronger than mine. I asked to see Adam's individual images to ascertain if my processing was the problem, and he very obligingly gave me links to them. Immediately the strength of his filter color signals showed through; I just did not have the color grasp that he did so comforted myself that 7,000 feet up was a big factor.

One subscriber put up a fantastic black and white image of the famous Horsehead Nebula. Praise came from every direction, and then someone spotted that it was identical, apart from some *Photoshop* modifications, to another member's image. It takes all sorts! It transpired that the originator of the image had spent £1000 on 24 hours of telescope time and 80 hours in processing the picture! He was understandably miffed!

Guy Hurst is the editor of *Astronomy* magazine, and coordinates "discovery" procedures for Europe. He issues regular bulletins concerning discoveries and vets reports of "new events" and is kept particularly busy by Tom Boles, Mark Armstrong, and Ron Arbour. Guy sends out reports of gamma ray bursts found by satellite where images are needed from amateurs to see if there is any optical counterpart to the glow. Speed is of the essence since the bursts are very short lived. Determined to do my part I sent in my first long exposure image of the sky that was defined. After a little delay

Guy acknowledged the image with the comment "Nice picture—wrong bit of sky!" I am pleased to say that subsequently I have made sure to image the right coordinates and have found optical events at the specified places.

I take my hat off to the guys who search for supernovae in this country with such so-so weather. They have infinite patience, amazing resilience, and very understanding spouses, so this is not for me. However, I am only too happy to take check images should these be required. In this, *The Astronomer* group, there are tales of expeditions to the most ungodly places to attempt to see phenomena, such as solar eclipses and meteor showers. One group took around 48 hours to get to a remote Pacific island, suffering all sorts of deprivations on the way, and saw nothing for their trouble. Another group in Libya saw a supposed 1-hour journey in a decrepit smelly bus take 6 hours before they became enmeshed in a traffic jam. They never got to the observing station and had to set up in the sand.

I will stand on the balcony for half an hour at times of meteor showers but am not greatly thrilled about the prospect of a few grains of space dust entering our atmosphere. As a young boy, my Uncle Tom gave me a meteorite almost as big as a fist. Now this really was something, and I was in awe about a body that had come from "out there." (He also gave me a one pound bank note, and I might have prized that even more.) Many years ago, in a bus, I saw a huge red fireball streak across the sky going north to south almost parallel with the horizon. Now, that got my attention! Later, after the astronomy bug had bitten, I saw a very bright meteor travel from Draco through Hercules, where it exploded, not far above the horizon, in a myriad of colors. It looked very close but I daresay was far away.

CHAPTER TEN

Photographic Results

It is important to retain a record of the details of any image that you take. Maybe an alien signal will arrive from a particular source and any historic image will be much prized! Yes, dream on! In reality, back images are often needed and will sometimes give a portrayal of events prior to discovery. This happens from time-to-time with supernovae.

In 2003 Jay McNeil imaged Messier 78, and in Lynds 1630 he noticed a nebula that he had not seen before. Indeed, it was a new nebula and thus it now carries Jay's name. (It is thought that the cloud was caused by a star that has intermittent outbursts, and these illuminate debris from previous actions.) Messier 78 is a glorious deep space target, and many images had been taken by a variety of people in the months before so that, with hindsight, it was possible to observe the creation of this nebula. I took an image 1 year before Jay; there might just have been the slightest suggestion of something about to happen at the relevant spot. It was a lesson to all that not only should you acquire the picture but also take a good look at what you have got.

Saving in FITS format means that a good file of information is attached to the image. Being somewhat pessimistic about the ability to retain electronic information I have devised my own means of recording what may well be the most important items: the name and date. I simply allocate a double-digit lettering system starting with "AA" in the title to the image. I keep a log with the date, type of camera and telescope in us, and the configuration of the telescope. I include notes about the weather and Moon. Very occasionally there is a cross denoting one of those rare splendid nights. I am now up to "ZD" so will soon have to embark on a three-letter system. Supernova hunters image many hundreds of galaxies in one night, so this may not be for them, but they must be sure to back-up their work.

I set out here images of some of the brighter objects with a short note regarding each of them.

Figure 10.1. NGC 206.

NGC 206 in Andromeda. Right ascension 00 hours 41 minutes
Declination plus 40 degrees 47 minutes. Distance about 2.5 million light years (Fig. 10.1).

RCOS at prime focus on September 15, 2004.

This was a moonless night of exceptional seeing. The target star cluster is within the great Andromeda Galaxy, and this was the first occasion on which I had acquired a sufficiently good signal to determine the color of individual stars beyond our own galaxy. Not only can the stars be seen to be mainly young bright blue stars, but hydrogen alpha regions are clearly visible in red. The images are shown here in black and white. The bold stars in this image are foreground stars in the Milky Way. The NGC206 stars comprise the group of smaller stars at the center of the image.

NGC 281 in Cassiopeia.

Right ascension 00 hours 53 minutes
Declination plus 40 degrees 47 minutes. Distance 9,500 light years.
Takahashi FS128 at prime focus on November 24, 2002.
This nebula is typical of a hydrogen alpha star forming region. The young hot star at the center is irradiating the surrounding molecular cloud and promoting the birth of new stars. Several dark Bok globules can be seen where star birth may be ongoing within them (Fig. 10.2).

NGC 869/884 in Perseus.

Right ascension 02 hours 53 minutes
Declination plus 57 degrees 10 minutes. Distance 7,300 light years.

Figure 10.2. NGC 281.

Takahashi FSQ November1, 2006. This pair of star clusters can easily be seen from a good site with the naked eye and are separated by just a few hundred light years (Fig. 10.3).

NGC 1499 in Perseus.

Right ascension 04 hours 04 minutes.

Declination plus 36 degrees 23 minutes. Distance 1,500 light years.

Takahashi FSQ 10th December 2005.

Part of the "California Nebula" this emission nebula shines in red as hydrogen atoms combine with electrons. This picture was one of a series of wide-angle images that I took in the autumn and early winter of 2005. The skies in the United Kingdom, though a little hazy, were remarkably stable during this period (Fig. 10.4).

NGC 2024 in Orion.

Right ascension 05 hours 42 minutes.

Declination minus 01 degree 50 minutes. Distance 1,500 light years.

Meade LX 200 operating at F6.3. December 4, 2002

The Flame Nebula, a dramatic emission nebula with prominent dark lanes.

On December 4, it was arranged that I would give a presentation to Oxford University Space Society (quite a different kettle of fish from the Oxfordshire League of the Women's Institute). Now, I am no astrophysicist, so you could say that I was somewhat apprehensive! Nevertheless they were very kind to me and no hand grenades were lobbed in my direction. On leaving Oxford the sky was awesome, with no Moon, so I spent the night imaging "The Flame." A bonus for me was that Patrick Moore included this image in the reprint of his *Atlas of the Universe* (Fig. 10.5).

Figure 10.3. NGC 869–884.

Figure 10.4. NGC 1499.

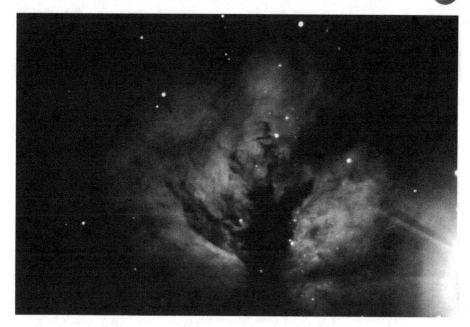

Figure 10.5. NGC 2024.

NGC2237 in Monoceros.

Right ascension 06 hours 31 minutes
Declination plus 05 degrees 02 minutes. Distance 5,500 light years.
Takahashi FSQ. January 11, 2006.
Red in color and known as the "Rosette," this is a spectacular emission nebula (Fig. 10.6).

NGC 2264 In Monoceros.

Right ascension 06 hours 41 minutes
Declination 09 degrees 53 minutes. Distance 2,500 light years.
Takahashi FSQ. December 27, 2005.
Known overall as the "Christmas Tree" nebula (here seen lying on its side), the peak of the tree is known as the "Cone" nebula or "Madonna and Child." The seeing for this image was again very good. There was no Moon, but as the exposure was through a hydrogen alpha (HA) filter moonlight would have been excluded (Fig. 10.7).

NGC 2903 in Leo.

Right ascension 09 hours 32 minutes.
Declination plus 21 degrees 28 minutes. Distance 20.5 million light years.
RCOS at prime focus, March 1, 2006.
From Earth we have a very good view of this face on barred spiral galaxy of similar structure to the Milky Way (Fig. 10.8).

NGC 4449 in Cannes Venatici.

Right ascension 12 hours 28 minutes.

Figure 10.6. NGC 2237.

Figure 10.7. NGC 2264.

Figure 10.8. NGC 2903.

Declination 44 degrees 03 minutes. Distance 10 million light years.
RCOS at prime focus. March 20, 2004.
This irregular galaxy might bear some comparison with our companion galaxy, the Large Magellanic Cloud (Fig. 10.9).

NGC 4565 in Coma Berenices.
Right ascension 12 hours 36 minutes.
Declination plus 25 degrees 45 minutes. Distance 30 million light years.
RCOS at prime focus. May 18, 2004.
For me this is the most dramatic of all the nearby edge-on galaxies. Missed by Monsieur Messier, it has a glorious central bulge crossed by striking dust lanes. It has a diameter of over 100,000 light years (Fig. 10.10).

NGC 4631 in Cannes Venatici.
Right ascension 12 hours 42 minutes.
Declination 32 degrees 30 minutes. Distance 27 million light years.
RCOS at prime focus. April 25, 2006.
This spiral galaxy is known as "The Whale" and has a small elliptical companion galaxy NGC 4627 beneath it (Fig. 10.11).

NGC 6946 in Cygnus.
Right ascension 20 degrees 35 minutes.
Declination plus 60 degrees 11 minutes. Distance 10 million light years.
RCOS at prime focus. August 31, 2004.

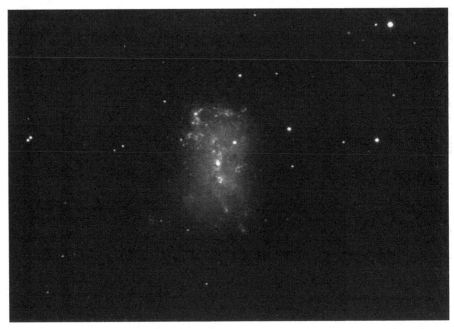

Figure 10.9. NGC 4449.

Figure 10.10. NGC 4565.

Figure 10.11. NGC 4631.

This nearby spiral galaxy is seen through dust of the Milky Way. There is a proliferation of star forming and supernovae regularly seen here: Since 1917 seven supernovae have been observed. This picture was taken on a night of excellent seeing (Fig. 10.12).

NGC 6946

RCOS operating at F6.75. October 18, 2004. Some two weeks after the taking of Fig. 10.12 a supernova was reported in this galaxy by Stefano Moretti and confirmed as SN2004et. Because of cloudy conditions I was unable to take a further image of the galaxy until October 18, and this image shows the interloper (the bright stars in the picture are foreground stars in our galaxy.) I subjected my initial image to close scrutiny but could detect no sign of any pre-discovery activity at the supernova site. Subsequent appraisal was that the exploding star was of massive size and that the event was likely to have given rise to the birth of a black hole (Fig. 10.13).

NGC6979 in Cygnus.

Right ascension 20 hours 51 minutes.

Declination plus 32 degrees 10 minutes. Distance 1,400 light years.

Takahashi FSQ. August 2, 2005

This supernova remnant is part of the large Veil Nebula or Cygnus Loop, a broken ring representing debris from a star that exploded some 5,000 years ago. At that time an observer would see a massively bright object in the sky of minus 8 magnitude, approximately the intensity of a crescent Moon (Fig. 10.14).

NGC 6992 in Cygnus.

Right ascension 20 hours 56 minutes.

Figure 10.12. NGC 6946 (2004–09–01).

Figure 10.13. NGC 6946 (2004–10–18).

Figure 10.14. NGC 6979.

Declination plus 31 degrees 41 minutes. Distance 1,400 light years.
RCOS operating at F6.75. October 11, 2004.
This shows a close up of the shock wave emanating from the exploding star (Fig. 10.15).

NGC 7000 in Cygnus.
Right ascension 20 hours 59 minutes.
Declination plus 44 degrees 32 minutes. Distance 2000 light years.
Takahashi FSQ. September 22, 2005.
The Pelican and North American nebulae form a fascinating area of sky, overburdened with activity. This picture comprises a mosaic of four images put together over two nights. To the left of the image is the Cygnus Wall, which is about 15 light years long and features a molecular cloud lit and eroded by bright young stars (Fig. 10.16).

NGC 7380 in Cassiopeia.
Right ascension 22 hours 47 minutes.
Declination plus 58 degrees 10 minutes. Distance 10,000 light years
RCOS operating at F6.75. October 25, 2004.
This is an H2 region supporting star growth and was imaged through a HA filter (Fig. 10.17).

NGC 7479 in Pegasus.
Right Ascension 23 hours 05 minutes.
Declination plus 12 degrees 21 minutes. Distance 105 million light years.
RCOS at prime focus. September 14, 2004.
This barred spiral galaxy has a width of some 160,000 light years (Fig. 10.18).

Figure 10.15. NGC 6992.

Figure 10.16. NGC 7000.

Figure 10.17. NGC 7380.

Figure 10.18. NGC 7479.

Barnard 169-71 in Cepheus.

Right ascension 21 hours 59 minutes
Declination plus 59 degrees 04 minutes.
Takahashi FSQ. November1, 2006.

Sometimes referred to as the "Fish," this dark nebula encompasses some bright stars. This image was taken at full Moon and the color version came out well, as the telescope was shaded by the dome. I took some HA-filtered information and was surprised to see that this added to the depth of the luminance image. This object will not radiate in HA, but perhaps the use of the HA filter blocked any reflection there might have been (Fig. 10.19).

Comet Linear (C/1999S4).

This comet was discovered by the Lincoln Near-Earth Asteroid Research program in New Mexico. It was imaged using the LX200 at prime focus by the method of superimposing five 1-minute images. It can be seen that the comet was scooting across the sky since the download time for each picture was 20 seconds. It was taken on July 19, 2000, in Ursa Major at a distance from Earth of about 34 million miles. A week after this image was taken the comet disintegrated on its last and probably only approach to the Sun. Cloudy conditions prevented me from capturing this event on camera (Fig. 10.20).

Comet Neat (C//2001 Q4).

Imaged low in the west on February 3, 2002. The distance to the comet was about 0.1 AU, and the orbital period is calculated to be around 37,000 years (Fig. 10.21).

Figure 10.19 B169–71.

Figure 10.20. Comet Linear.

Figure 10.21. Comet 2002VI (Neat).

IC 342 in Camelopardalis.

Right ascension 3 hours 47 minutes.

Declination plus 68 degrees 07 minutes. Distance 6.5 million light years.

RCOS operating at F6.75. Image taken on November 4, 2004.

This face-on spiral galaxy would appear much brighter but for the extinction of light due to its alignment with the plane of the Milky Way (Fig. 10.22).

IC410 in Auriga.

Right ascension 05 hours 23 minutes.

Declination plus 33 degrees 25 minutes. Distance 12,000 light years

RCOS operating at F6.75. January 2, 2005.

Each of the "tadpoles" has a length of around 10 light years (Fig. 10.23).

IC434 in Orion.

Right ascension 05 hours 41 minutes.

Declination minus 02 degrees 26 minutes. Distance 1,500 light years.

Takahashi FSQ. May 5, 2006 through a hydrogen alpha (HA) filter.

This spectacular area centers on the emission nebula IC434 and features the "Horse-head" dark nebula Barnard 33, which is about 5 light years long (Fig. 10.24).

Barnard 33 The Horsehead Nebula

RCOS operating at F6.75. January 2, 2005.

B33 is hard to discern through a telescope but is quickly apparent through a CCD camera (Fig. 10.25).

Figure 10.22. IC 342.

Figure 10.23. IC 410.

Figure 10.24. IC 434 – B33.

Figure 10.25. IC 434.

IC1318 in Cepheus.
Right ascension 20 hours 20 minutes.
Declination 40 degrees 16 minutes. Distance 3,700 light years.
Takahashi FSQ. Image acquired on August 8, 2005.
A hydrogen alpha exposure formed the luminance for this image (Fig. 10.26).

IC 1396 in Cepheus.
Right ascension 21 hours 39 minutes
Declination plus 57 degrees 32 minutes. Distance 2,450 light years.
Takahashi FSQ. Taken on August 10, 2005 (Fig. 10.27).

IC 5146 in Cygnus.
Right ascension 21 hours 53 minutes.
Declination plus 47 degrees 18 minutes. Distance 10,000 light years.
LX200 operating at F6.3. Image taken on October 12, 2002.
This emission and reflection nebula is the well-known Cocoon Nebula (Fig. 10.28).

Messier 13 in Hercules.
Right ascension 16 hours 42 minutes.
Declination plus 36 degrees 26 minutes. Distance 23,000 light years.
RCOS at prime focus. Image taken on July 11, 2006.
Known as the Great Globular Cluster in Hercules it contains several hundred thousand stars and is a showpiece object through a telescope; the brightest cluster in the northern sky (Fig. 10.29).

Figure 10.26. IC 1318.

Figure 10.27. IC 1396.

Figure 10.28. IC 5146.

Figure 10.29. Messier 13.

Messier 16 in Serpens Cauda.

Right ascension 18 hours 19 minutes.

Declination minus 13 degrees 47 minutes. Distance 7,000 light years.

RCOS at prime focus. Image taken around July 16, 2006.

These are the famous star birth "Pillars of Creation." Being low in the sky for me, I had to rather dodge around the trees, and it took several days to build the picture. The image is a combination of hydrogen alpha (HA)-filtered and -unfiltered exposures (Fig. 10.30).

Messier 20 in Sagittarius.

Right ascension 18 hours 02 minutes.

Declination minus 23 degrees 02 minutes. Distance 2,200 light years.

RCOS operating at F6.75. Image taken on June 24, 2003.

This is the famous Trifid Nebula and was my first target with the RCOS (Fig. 10.31)

Messier 27 in Vulpecula.

Right ascension 19 hours 59 minutes.

Declination 22 degrees 44 minutes. Distance 1,250 light years.

RCOS at prime focus. Image taken on June 14, 2004.

The famous Dumbell planetary emission nebula, the type of remnant that the Sun will leave when nuclear fusion ceases in its core (Fig. 10.32).

Messier 33 in Triangulum.

Right ascension 01 hours 34 minutes

Declination 30 degrees 42 minutes. Distance 3 million light years.

Figure 10.30. Messier 16.

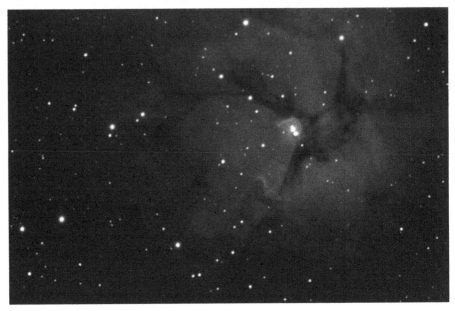

Figure 10.31. Messier 20.

Figure 10.32. Messier 27.

LX200 operating at F6.3. Image taken on January 20, 2003.

This nearby member of the Local Group of galaxies provides a close up look at a spiral galaxy. The day preceding the night of this picture was stormy; as the heavens cleared the sky became exceptionally transparent and surprisingly stable. I rapidly secured 30 minutes of luminance and 10 minutes of red, green, and blue, but then the clouds returned for the night. The conditions were so good that I had determined to "image my socks off," but yet again Mr. Frustration came along (Fig. 10.33).

Messier 42 The Orion Nebula.

Right ascension 05 hours 35 minutes.

Declination 05 degrees 23 minutes. Distance 1,500 light years.

LX200 operating at F6.3 on February 3, 2003.

The magnificent Orion Nebula offers a considerable challenge to astro-imagers due to the great dynamic range across the field. For this image (taken in color) I combined a series of exposures at 1, 10, 60, and 300 seconds. Each of these was a combination of images so that the total number taken was well over a hundred (Fig. 10.34).

Messier 63 in Cannes Venatici.

Right ascension 13 hours 16 minutes.

Declination plus 41 degrees 59 minutes. Distance 37 million light years.

RCOS at prime focus. Image taken on April 16, 2005.

The Sunflower Galaxy. This image was taken on a night of exceptional seeing (Fig. 10.35).

Figure 10.33. Messier 33.

Figure 10.34. Messier 42.

Figure 10.35. Messier 63.

Messier 64 in Coma Berenices.

Right ascension 12 hours 57 minutes.

Declination plus 21 degrees 38 minutes. Distance 17 million light years.

RCOS at prime focus. Image taken on April 16, 2005.

This picture of the Black Eye Galaxy was taken on another night of exceptional seeing (Fig. 10.36).

Messier 82 in Ursa Major.

Right ascension 9 hours 56 minutes.

Declination plus 69 degrees 40 minutes. Distance 12 million light years.

RCOS at prime focus. The basic image was taken on March 4, 2004, and 3 1/2 hours of HA information was acquired on April 7 and 8, 2007. This AGN (active galactic nucleus) galaxy was probably disturbed by the passing of the nearby galaxy Messier 81 about 600 million years ago (Fig. 10.37).

Quasar 957 + 561 in Ursa Major.

Right ascension 10 hours 02 minutes.

Declination plus 55 degrees 54 minutes. Distance 9.1 billion light years.

RCOS operating at prime focus. Image taken January 29, 2006.

I have difficulty in comprehending that this split source of light has been traveling for over 9 billion years before striking my mirrors. At a redshift of 1.41 the light from this quasar has been enhanced by gravitational lensing caused by a massive intervening galactic cluster at a distance of 3.9 billion light years (redshift .36). The cluster divides, magnifies, and bends the light in such a way that the light at Point A, having traveled the shorter distance, arrives 417 days before the light at Point B (Fig. 10.38).

Figure 10.36. Messier 64.

Figure 10.37. Messier 82.

Figure 10.38. Q957+561.

Figure 10.39. SH2–155.

Sharpless 155 in Cepheus.

Right ascension 22 hours 57 minutes.

Declination plus 62 degrees 30 minutes.

Takahashi FSQ. Image taken on August 21, 2006.

This image of the Cave Nebula is a combination of hydrogen alpha (HA)-filtered exposures (Fig. 10.39).

CHAPTER ELEVEN

Stargate 4173 at Grimaldi Tower

Through the SBIG camera users group I have, from time-to-time, made suggestions to people wishing for views about different styles of observatories. In particular, I became involved with Bob Antol in discussing the plans he had for a grand observatory north of New York. It was to be the Grimaldi Observatory, named after his prize cat. Now, my wife has cats and one of them, Monty, a big male kitten, has a bad habit of running up your leg. Fresh off the tennis court, in shorts, he carried out this act on me—ouch! He also has the habit of walking across the keyboard while an image is downloading. My observatory would not be named after a cat! One of the joys of astronomy is the manner in which participants pull together in helping one another, and it was a pleasure to give Bob some of the pros and cons of building the Crendon Observatory.

Three years ago I had no junk-mail filter and was being inundated with the wretched stuff: one hundred a day at least. There was a mail entitled "man from Wyoming" and my finger was on the delete button when something told me to pause. It was from a man called Brad Meade who was coming to England in a few days and was thinking of building an observatory adjoining his house. He had seen my Web site and wondered if he could stop in. Of course, he was most welcome. He arrived with his delightful wife Kate and boys Sam and Tucker. After the conducted tour they did us proud at the local hostelry. Margaret and I know one other resident of Wyoming and, would you believe, Brad is on the same Board of Directors as Alan!

Brad has to be a most trusting individual, because he duly built his observatory taking my recommendation about the dome, the telescope, and the mount. Brad tells me he has just purchased a vineyard with a view to constructing a distillery. "Brad's Mead" comes to mind as a title that might be appropriate for the hooch.

Following now is an account of Bob Antol's efforts concerning the construction of his observatory (Fig. 11.1). In the next chapter is Brad Meade's account of how he built his observatory.

Figure 11.1. Grimaldi observatory.

41 degrees 36 minutes 49 seconds North, 73 degrees 40 minutes 16 seconds West
Robert A. Antol.

In the Beginning

My fascination with astronomy and observing the heavens started when I was 10 years old. I am 50 years old now, so astronomy has been a part of my life for 40 years! Of course, life was much simpler in the early 1960s. An observing session back then was as simple as going into the backyard when the skies were clear and looking up.

When my parents bought my brother and me our first telescope, we couldn't believe the stark beauty of the Moon, the magnificence of the planets, and the wonders of the universe. An observing session with this first primitive telescope consisted of taking the rather lightweight tube and tripod assembly on an alt-azimuth mount out into the backyard and finding a suitable location to view an object such that it was not blocked by any trees. If we wanted to see something that was behind a tree, we picked up the telescope and moved to another location in the yard.

As I got older, the telescopes, tripods, and mounts got bigger and heavier and the accessories more numerous. An observing session now had to include the set-up and take-down phases for the tripod and telescope. The days of easily going into the backyard and observing for a few minutes were gone.

A typical session started out by hauling the 19-pound heavy duty variable-height Meade standard field tripod to the location in the yard where the observing would be performed. Then, I would go back inside the house and bring out my suitcase of accessories. If I was just observing, a single trip would suffice. But, if I was going to connect my camera, then a couple of trips were necessary to ensure I had everything I needed.

There was a final trip back to the house to haul out the 33-pound Meade LX-50 8-inch Schmidt–Cassegrain telescope. Before mounting the telescope on the tripod though, I would first level the tripod head and polar align the tripod to the best of my ability. I would also set the tripod height to a comfortable viewing level. When satisfied, I could lift the telescope into place and lock it onto the tripod.

The telescope's built-in bubble level was very helpful in fine-tuning the tripod legs so the telescope was perfectly level. Polar aligning was then performed and when all was perfect, observing could then begin.

The total elapsed time for the set-up procedure was about 15 minutes, and an equal amount of time was needed for the reverse process of taking everything down and stowing it away. This meant a 5-minute observing session would require a total of 35 minutes of time including set-up and take-down, so it was hard to justify a short observation session.

There had to be a better way. I had always thought that a permanent mounted polar-aligned telescope in an outdoor building would be the most convenient thing to have. Researching this idea would eventually lead to the construction of the Stargate 4173 at Grimaldi Tower Observatory.

Use a Roll-Off Roof?

Prior to deciding on a dome, however, I seriously considered building a roll-off roof observatory. A roll-off roof is basically a small shed with a removable roof. In actuality, the roof is not removed but is rolled off to the side on rails connected to the shed structure. My initial plans called for a one and a half story building. The bottom floor would be considered the "warm room" or the "control center." I imagined myself walking down the stairs after several hours of observing in the cold and entering a room that housed a microwave oven. Making a piping hot mug of cocoa and sitting in a soft recliner would help take off the chill before I walked back up the stairs into the cold evening for continued viewing.

This was an interesting idea, but I began running into problems as I sketched out initial designs on paper. My desire to have a refrigerator, microwave oven, and a soft recliner dictated the size of the room. For mounting the telescope, I had envisioned a solid concrete pier shooting through the center of the first floor room. The pier, in addition with everything else I required, forced the room to be larger. A larger room dictated a larger building footprint, which resulted in a larger roof. As the roof got larger, it got heavier. I then had to devise a plan to roll a very heavy roof onto the side rails. Plus, as the roof got heavier, the supports for the side rails got more massive. What was this final structure going to look like?

The Dome

With the many unanswered questions regarding the one and a half story roll-off roof observatory in the back of my mind, I happened one day to see an Ash Dome observatory advertisement in *Sky & Telescope* magazine. I was very impressed with the overall look of the dome. I have always liked observatory domes; but never thought that I would actually own one. With the realization that personal-sized domes were available, I began rethinking my plans regarding the roll-off roof. The first task to be completed was to test the waters with my wife, Barb, to see if a dome in the yard was a possibility.

As it turned out, my wife was very receptive to the idea of replacing the roll-off roof design with the dome. I showed her the Ash Dome advertisement. She liked the classic look and thought the idea was really cool. Fortunately for me, "convincing the better half" was a non-issue. For anyone else thinking of building an observatory, though, you definitely need to have your significant other involved and in agreement with the plans for your structure.

Visiting an Ash Dome

I sent off a request to Ash Dome for information regarding the dome: what sizes were available, what were the costs, shipment details, etc. Shortly thereafter, I received an e-mail with answers to all of my questions, including a list of nearby Ash Dome owners along with their dome locations. To my delight and surprise, one of the local colleges had three Ash Domes of different sizes.

I called the astronomy department at the college. They told me free observing sessions were held every clear Wednesday evening. I thanked them, but told them I wasn't interested in looking through their telescope at night, but rather wanted to come see their domes in the daylight. At first, they thought this was a strange request, but I told them of my plans for designing and building an observatory, which included the incorporation of an Ash Dome. The astronomy professor understood and agreed to have one of his grad students provide us with a private tour of the domes.

We drove to the college and met with the grad student, who gave us a tour of each of the three domes; one was manually operated while the other two were motorized. The Ash Dome was amazing! There is nothing quite like standing inside the observatory and seeing the actual dome. It is very impressive. For anyone interested in building an observatory, I recommend being able to see, touch, and experience the dome you plan to purchase ahead of time.

In addition to just "seeing" the domes, we also received some valuable information on the "workings" of them. One thing we learned was the importance of ensuring that the lower aperture door of the dome, when opened, would have sufficient clearance for proper 360 degree rotation to avoid obstacles. A design flaw with the college building required them to have to raise the lower aperture door to avoid collision with the entrance way structure. I believe the most important aspect, though, was getting the actual feel for how much room was available inside the different-size dome structures.

Concept

After seeing the Ash Dome in person, we were convinced the dome was the observatory type for our yard. The roll-off roof design would have been nice, but we were both happier with the dome. The next step was to select the location in the yard for the dome and decide if the observatory would be ground level or one and a half stories—similar to our roll-off roof plans.

Our house is a two-story colonial situated on a hill. There was one ideal location in our backyard for the observatory, but if selected, we would have had to sacrifice a beautiful tree. Preserving all of the trees meant selecting a location that was not entirely suitable for astronomical observing. We knew what we wanted, but we were having a difficult time in arriving at a mutually acceptable location for the structure. As we continued to think this through, an idea surfaced for mounting the dome on our existing garage. This was promising in that the observatory gained height and reduced obstructions; plus, no trees would be lost.

We also thought it would be extremely convenient to be able to enter the dome without having to go outside. This attraction convinced us to abandon the stand-alone observatory concept. We wanted to devise a plan that incorporated the dome into our house.

One problem with building the observatory atop the garage was losing a car space as a result of the telescope pier. It would be possible to build the Ash Dome on top of the garage without the pier, but the vibrations to the telescope would make the observatory useless (at least from an imaging point of view).

Design

Since the pier was a definite requirement, we started to search for other alternatives. The next idea in the evolution of our observatory design was to extend the back of the garage sufficiently to accommodate the pier without interfering with the interior of the existing garage and then mounting the dome above that central spire. This idea had real promise. So we began sketching out some pictures of the back of our house with an attached dome.

We knew there were still many architectural issues to be addressed, but we also knew we were getting closer to our final design. Prior to deciding on the extension behind the garage, we walked around our house trying to envision what it would look like if we added the observatory to different sides of the house. We also drew a sample floor plan layout of this proposed extended garage, depicting what we had in mind.

After continued discussions, we arrived at the preliminary design that would form the basis of our observatory—a full room extension behind the garage with an attached tower and the dome mounted atop the tower. We still ended up sacrificing that one tree, but the many benefits of the design outweighed that loss.

Additional Research

The Internet is a fantastic place to find other Ash Dome owners. When I found someone who had been through the process of ordering, assembling, and eventually using the dome, I would contact him or her and ask many different questions. Almost everyone I contacted responded with helpful information. A very common question I asked was, "If you could do it all over again, what would you do different?" Their responses definitely helped me in revising and finalizing my own plans.

One very useful resource during my early research was the Ash Dome Web site (http://www.ashdome.com/page5.html). This site has pictures and Web site links of what other Ash Dome owners have done. I discovered one dome, in particular, that impressed me since it was mounted atop the house similar to what I intended to do. That dome of interest was the Crendon Observatory owned by Gordon Rogers and located in the United Kingdom.

After visiting Gordon's Web site, the immensity of the project I was about to undertake really hit home. There were many unanswered questions I still had, and I wondered if my efforts would result in an observatory as beautiful as the Crendon. The only way to get some of my questions answered was to contact Gordon.

He was extremely helpful and answered all of my questions. I e-mailed him several times with different queries. In many cases, his responses prompted me to rethink some aspects of my proposed design. This, in turn, resulted in even more questions. Some of the answers he provided to me played a crucial role in the finalization of my own plans. For anyone considering building a first-class domed observatory, here is a list of questions to consider (similar to what I asked Gordon):

- If you could do it all over, would you still purchase the Ash Dome?
- Did you either heat or air condition your observatory?
- Are you happy with the size of the dome you purchased? If you could do it over again, would you go with a smaller, same size, or larger dome?
- What is the height of the walls?
- How do you gain access to the observatory?
- What kind of pier do you use?
- If the pier is concrete, how is telescope stability?
- Do you have any additional information or helpful hints that might aid me in the design of my observatory?

The Town Code—Know It Ahead of Time!

Every state, every town, every city, every village has their own unique set of rules associated with building on land you own. These rules are in place to protect you—and your neighbor. The construction of a domed observatory is unique enough to warrant a pre-investigation into the town code so you will know whether or not you have the right to build. Check with your town hall on the appropriate steps you need to take in order to get the proper permissions.

I live in the state of New York; the county of Dutchess; the town of Beekman; the hamlet of Poughquag. In Beekman, there is an official town zoning code available online for anyone to review. The recommended steps for any construction project include:

- Draw a sketch of your project and get pre-approval to continue.
- Work with a licensed architect to develop detailed construction plans.
- Obtain official town approval and building permits.

Plans (Architectural)

Our town would not grant a building permit for a project of this size without the submission of a licensed architect's drawings. The phone book was a good place to start the search. But, not surprising, we didn't see any architect who advertised that they specialized in the design of home observatories. However, we were fortunate in that a friend from work recommended an architect they had worked with on some home improvements and were happy with the results. We set up an appointment with that architect so we could explain what we had in mind.

Prior to our initial meeting with the architect, another friend gave us some more good advice. Whenever you work with an architect, be sure to get what you want—not what the architect "thinks" you want. In the end, we were very lucky in the selection of our architect. He not only gave us what we wanted, but provided suggestions and improvements that we didn't think of.

Our architect was Dan Contelmo from Architechniques, now doing business as Millbrook Architects. For our initial meeting with Dan, I had the brochure from Ash-Dome as well as some hand-drawn sketches of our ideas. We walked around the yard and showed Dan where we wanted the new addition. I then provided the requirements for the dome. My initial requirements ended up driving both Dan and I crazy. I wanted a 12-foot diameter dome, a full-size door to enter the observatory room, and the wall height was limited to 5 feet 2 inches. I explained to Dan the line of reasoning behind this decision. If you had standard 8 foot high walls and you wanted to look at something near the horizon, you would not want to climb a step ladder in order to peer through the eyepiece. The telescope height had to be comfortable for viewing all portions of the sky. With my list of requirements in hand, Dan returned to his office to design his preliminary sketches.

When we next met with Dan a few weeks later, he had several different options for gaining access to the observatory room given the parameters he had to work with. The first sketch had a submarine-type hatch with access via a ladder. This provided the maximum amount of usable floor area in the observatory, but we rejected this idea immediately because we knew we would have difficulty gaining access to the room as we got older. Plus, it would be very difficult to haul the telescope up the ladder and through the hatch. Finally, not all visitors to our observatory would be able to comfortably enter and exit the observatory via the ladder.

The next sketch had a full-size door that would open into a pit in the room. A set of stairs would bring you out of the pit into the observatory. This gave easy access to the room for anyone of any age. But, a full-size door is close to 3 feet in width. Carving out a pit in the observatory room to accommodate the door plus a railing (so you wouldn't fall into the pit) took up too much space. Considering we only had a 6-foot radius to work with, this plan left very little room next to the telescope.

Another sketch was a twist on the previous concept. Once you were in the observatory and had the door closed, a drop down panel would be lowered covering the pit. This would provide maximum floor space, but also trapped you in the observatory. It would be awkward for additional people to enter once the panel was in place.

After seeing the difficulties Dan was having, I realized some problems would go away if I relaxed some of the initial requirements. I didn't want to increase the wall height, so I checked the price list in the Ash Dome brochure to see how much more it would cost to increase the dome diameter from 12–16 feet. The difference in price was manageable, so I asked Dan if it would help if we went to a 16-foot dome. His eyes lit up with this change, and he immediately began sketching some new possibilities.

Barb looked at my 8-inch telescope and commented, "Isn't a 16-foot dome rather large for the 8-inch telescope?" I calmly replied, "Yes, so I'll have to buy a larger telescope!" It was finally out in the open. My ulterior motive was finally exposed. There is a lesson to be learned here: If you want a larger telescope, convince your spouse that you need a larger diameter observatory dome first. The newer and improved larger telescope will naturally follow.

With the new parameters to work with, Dan came back with a wonderful solution. Originally, the lower room of the tower was only going to be accessible from the outside and would be used as a storage room. Original plans also called for a door set into the far wall of the rectangular addition to open onto a stairway that would ascend to the observatory. The switch to a 16-foot dome dramatically altered these plans—for the better!

Dan's new plans resulted in merging the octagonal room (i.e., the lower room of the tower) with the rectangular addition. No longer would the octagonal room be considered a storage room only accessible from the outside. Now, this uniquely shaped room would be part of our living quarters. Entrance to the second floor observatory would be through a door in the lower octagonal room. The purpose of the door was to act as a temperature isolation barrier separating the observatory from the house. The stairs to the observatory would hug the inside four walls of the tower, eventually bringing you almost 180 degrees opposite the starting point. This design, incorporated with the 16-foot dome, provided plenty of space around the telescope.

It was essential that the pier be totally isolated from the building structure so this was one very important item I wanted to see very clearly stated on the plans. Dan accommodated this request by adding this:

Note:
Do not bear or connect any framing members on concrete pier.
Provide an isolation joint between the concrete pier and the slab.

Town Approvals

With the architectural plans complete, it was now time to get the town's approval and to obtain the building permit. I dropped off two copies of the plans with the town and asked them to contact me if there were any questions. The first question they asked was if I had a builder yet. I was hoping to get approval before selecting a builder. However, the town would not review and approve the plans and issue a building permit without having a builder selected. I also had to provide proof that the selected builder had sufficient insurance to cover the possibility of any "on the job" accidents. They didn't seem to have any other concerns, so I was encouraged by that. Approval appeared to be just around the corner. I couldn't believe how smooth everything was going so far.

But, as you will see, the seemingly smooth road ahead would have some curves and bumps.

Selecting the Builder

Just as architects don't advertise a specialty in designing observatories, builders don't either. So, I asked our architect to recommend a list of builders. Since the architect works on a daily basis designing homes and doing home improvements, it was natural to assume they knew who had the skills to tackle the job that I was pursuing. I was provided with four possible candidates. It was time to begin the interview process.

All four builders were contacted by phone. Three of the builders returned my calls. The fourth builder was knocked out of the bids right away since he didn't respond. Not returning a phone call from a prospective client just isn't indicative of good business practice.

Individually, the remaining three builders came over to the house, and I showed them the architect's drawings and the location in the yard for the construction. They all took their copies of the plans with them and indicated to me that it would take one to two weeks to analyze and determine a price for the project.

During one of our early meetings with the architect, we asked if there was a formula that could be used to estimate the cost of the project—sort of a "ballpark figure." He told us a certain dollar value per square foot based on the building materials we had in our plans. We computed the number and it turned out right about what we had intended to spend. Things looked pretty good. If the actual bids were close to that ballpark figure, we would be set.

Within a week, the first builder called with his bid. The bid he provided was 25 percent higher than what we had budgeted based on the architect's estimate. We were devastated. My dream of having a first-class domed observatory attached to a two-story tower incorporated into a new addition for our home was looking as if it would be just a dream. In order to afford this project, we were going to have to scale back some of the features to bring the price more in line with our budget. Would all three of the builders come in with bids this high? We would have to wait for the next bid.

A week later, the second builder called us. When he told us his bid, our spirits lifted. His bid was right where we thought it would be; right where we hoped it would be. This builder was Ed Peterson of Peterson Home Improvement.

The third builder never called us back with a bid. We started with four prospective builders; talked with three of them; and heard back from only two. One bid was exorbitant, while the other was reasonable. Various friends who had recently dealt with contractors warned us to be prepared to spend between 10 and 30 percent more than the original quoted price due to possible unforeseen costs. We asked Ed directly if the price he quoted was the price we would pay or if this number could change. Ed gave us his word that the price quoted was the price we would pay, barring any changes we introduced after contract signatures.

As it would turn out, Ed was true to his word. The quoted price was indeed what we paid and the workmanship was excellent. Our decision to work with Ed and his crew was a good one and one we did not regret.

Now that we had a builder, it was time to re-approach the town for final approval.

Getting Neighborhood Approvals

Unfortunately, approval was not immediate. I went down to the town hall and told them I had everything they requested and was ready to pick up the building permit. Not so fast, I was told. The town informed me I needed to present my plans to the Architecture Review Board (ARB). This was news to me. I questioned them as to why I wasn't informed of this esteemed body when I first presented my plans. The response I got back was simple—this board was newly created! I kept my composure and asked specifically which codes were in violation and why these plans needed to go before this ARB. They told me there were no specific violations, but because my planned addition was "different," further review along with neighbor approval would be required.

I pointed out that my observatory would look just like a silo. We live in a rural area and, fortunately for me, there is at least one silo within a half mile of the town hall. They acknowledged this was in my favor and indicated the review board would likely not have any issues. Therefore, I asked if it would be acceptable to submit written approval from my neighbors in lieu of having them attend a town meeting. They agreed that would be sufficient, and if I submitted the written consent from my neighbors, I would not have to go before the ARB.

I went home and created a single page document with the architect's depiction of the final project and the following text:

We (our name and address) have plans to build an addition onto the back of our house and will be abiding by all zoning regulations. The addition will be attached to the back of the garage and will extend into our backyard. We will also be incorporating an astronomical observatory built on an octagonal structure located on the Southwest corner of the new addition. The architect's proposed views of this project are included below.

Your signature on this note indicates that you concur with these plans and have no objections or concerns.

I printed up the correct number of copies and then went door to door to meet with every neighbor whose property was adjacent to ours. I obtained the five required signatures without any difficulty at all. As a matter of fact, our neighbors were all very interested in the project and couldn't wait to see it completed.

The very next day, I took the five signed letters down to the town hall. That did it! The town was ready to issue me my building permit to begin construction on my observatory. I informed Ed I had the approval, and he gave me a date as to when his crew could begin construction.

Construction

The actual start date for the project arrived pretty quickly. The crew assembled their equipment and supplies in the backyard and began work on the 4-month project. Chalk lines were drawn on the existing house foundation and on the grass showing the footprint of what would be my dream observatory. Up to this point, we only had artistic depictions of the project on paper in an architect's line drawings. Now we were able to actually see how much of the yard was going to disappear under the outline of the new addition.

The Foundation

A back-hoe was used for the job of neatly digging a trench whose depth would extend below the frost line. This trench would form the foundation of the new construction. The design of this project called for a 16-foot diameter octagon attached to the corner of a rectangle. This was easy to draw with an architect's CAD (Computer Aided Design) program, but digging 135-degree angle trenches using a back-hoe was no simple matter. Tom, the mason who was in charge of laying the cinder block foundation, was also the back-hoe pilot. He was an expert who commanded the machine to perform the delicate maneuvers needed to match the architect's plans.

Upon completion of digging the intricate trench system, the tedious job of laying hundreds of cinder blocks was begun. It took a total of two weeks for the blocks to be properly positioned along with the pouring of the concrete sub-flooring. If it seems that two weeks is an inordinate amount of time, you have to realize that the blocks at the corners of the octagon had to be cut at 45-degree angles with a diamond blade (see Fig. 11.2).

Dome Delivery

In the meantime, Ash-Dome was building my dome at their manufacturing facilities in Plainfield, Illinois. Ash-Dome assembled the dome at their plant and made sure all of the holes lined up, all of the pieces fit together properly, all of the electrical motors were operational, etc. When they were satisfied the dome "worked," they took it apart and packed it on a large 16 foot long by 6 foot high by 5½ foot wide pallet that weighed 2,800 pounds. This pallet was then loaded onto an 18 wheel semi-truck for its cross-country trek to be delivered to my home in New York.

Figure 11.2. The foundation footprint of the addition for the observatory.

The trucking company only provided roadside delivery to the base of my driveway, so I wondered how I was going to get the dome off of the semi and placed in my backyard near the construction site. I posed this question to Ash Dome, and they were able to offer a solution, since many of their customers faced this similar problem. The solution was actually quite elegant in its simplicity; it employed the use of a flat-bed tow truck. I asked the local garage if I could hire their flat-bed tow truck for an hour. When the semi carrying the observatory pulled up in front of my house, I phoned the garage and told them I was ready for them. The flat-bed came to our house, backed up behind the semi and raised its bed to the same height as the trailer. Then, the tow cables were strapped to the 2,800-pound pallet. The flat-bed driver activated the winch to begin pulling the observatory pallet out of the semi onto the back of the flat-bed.

Once the pallet was completely on the back of the flat-bed, the tow truck with the dome package aboard was driven to the back corner of the yard. The tow truck then angled the flat-bed and very carefully lowered the pallet onto the ground. Fortunately for me, the back-hoe that was digging the foundation of the addition was at work in the yard. The back-hoe attached a chain to the pallet and assisted the tow truck in gently lowering the pallet. Without the back-hoe's help, the pallet would have dropped to the ground a little harder, since it was basically sliding off the back of the tow truck. Without incident, the dome was safely tucked away in the corner of the backyard, waiting for the day when construction of the tower was complete and assembly of the dome could begin.

Figure 11.3. Arrival of the dome package.

Figure 11.3 shows the dome package arriving from Ash-Dome by way of an 18 wheel semi truck. A flat-bed tow truck pulls the pallet off the semi for transport to the backyard construction site.

Installing the Pier

One of the aspects that made the project so unique was the 2-foot diameter concrete pier that could not touch any part of the structure. Tom devised a simple but very effective way to construct this pier. Instead of building a custom form, he used a road culvert with very high crush strength. This was important as the culvert would be placed vertically on end and filled with concrete. It was vital that the culvert did not crack or break open during the concrete fill process.

The installation of the pier began with the digging of a square hole that was 5½ feet deep and located in the center of the octagon foundation. In this hole, concrete was poured to form a slab that was 5 feet long by 5 feet wide by 1¼ feet in depth. There was a cross-hatching of #4 rebar every 6 inches, positioned 5 inches above the bottom of the slab. Sticking out of the center of the slab were four #5 vertical rebar with a minimum 16-inch bend embedded in the slab. The imperial definition of a #4 rebar is.5 inches in diameter while the #5 rebar is.625 inches.

Figure 11.4. The rebar inside culvert pipe.

The four vertical rebar required a minimum of 4 inches from the outside of the concrete pier. These vertical rebar were attached with #3 horizontal rebar separated every
16 inches. A #3 rebar is .375 inches in diameter. After the slab cured, more vertical rebar
were added for height, with a minimum 3 feet overlap between sections (Fig. 11.4).

Once the internal skeleton was constructed, the road culvert was lifted by the back
hoe and centered in place over the rebar. Three wooden beams were then attached to
the outside of the culvert and screwed to stakes buried deep in the ground. Tom very
carefully ensured the culvert was in the center of the octagon, perfectly vertical and
securely braced (Fig. 11.5).

Scaffolding was then constructed around the culvert, allowing one to see down the
center from the top. Everything was in place for the fill process to commence.

Tom constructed a ramp next to the pier so the back-hoe would have sufficient
height for his plans. When the cement mixer truck arrived, he attached a hopper to
the end of the back-hoe. One of his helpers was positioned at the top of the scaffolding while the second helper lingered near the base of the pier with a rubber mallet.
The back-hoe swung the empty hopper over to the truck to be filled. After the hopper
was filled, the back-hoe very carefully lifted it above the top of the pier where the
lever was engaged to dump the concrete down the culvert shaft. While the back-hoe
returned the hopper to be refilled, the helper with the rubber mallet would continuously pound on the side of the culvert to remove any trapped air bubbles in the concrete (Fig. 11.6).

Figure 11.5. The pier tube prior to insertion of concrete.

Figure 11.6. The back-hoe delivers concrete to pier by way of hopper.

Since the pier was resting on the slab 4¼ feet below ground and rose through the first floor, which had 10-foot-high ceilings and extended through the floor of the observatory, the culvert pipe used was over 17 feet in total length. To fill the tube, it took Tom close to two dozen repetitive trips from cement mixer to pier top and back.

When the culvert was topped off, we all had to work fast because it was critical for the last step to be done correctly. I requested the design of the concrete pier to neck down from 2 feet diameter to 1 foot diameter for the top 2 feet. This was accomplished by placing a 2 foot long by 1 foot diameter Sonotube atop the filled culvert pipe. The Sonotube was then hand filled with concrete, smoothed, and leveled. The next task was to mount a steel plate with 3 J-bars into the Sonotube. The steel plate had to be perfectly level, and it had to point at Polaris. When finished, a Pier-Tech 2 telescoping pier would be bolted to this steel plate. However, details for this part of the story will come later.

Here is how I managed to point a steel plate at Polaris in the middle of the day. During one of the previous clear nights, I took a compass with a rotating bezel and pointed it at Polaris. I then spun the compass so the magnetic needle lined up with true north. This was a simple way to know exactly where Polaris was during the day. Standing atop the scaffolding, all I had to do was abut the compass (which had a flat side to it) against the steel plate and point the needle to magnetic north. Adjusting the plate in the wet concrete allowed me to pre-position the plate correctly. A bubble level was used to level the plate. When the concrete cured, I would have an extremely level, stable, and polar-aligned plate ready to accept the Pier-Tech 2.

Building the Dome

The construction of the addition went without incident (Fig. 11.7 the interior of the new addition with tower and pier in the background). After the octagonal tower was constructed and after the roof was in place on the addition, it was time to construct the dome. I volunteered to take vacation from work to help the carpenters assemble the dome. This was a once in a life-time project and I wanted to be a part of it. Fortunately, Ed did not have any problems with allowing me to help. And since this was volunteer work, I was not on Ed's payroll.

The 16 foot Ash Dome took a total of 5 days to build. There were three of us working on this, but a fourth person was used during some of the steps. Here is a breakdown of the 5 days of construction:

Day 1: Dome skirt
Day 2: Dome panels
Day 3: Rails
Day 4: Shutter doors
Day 5: Motors and control boxes

The Ash-Dome came with a 56-page assembly instruction manual—11 pages of text and 45 pages of drawings. I made additional copies of this manual for everyone involved with the assembly. I deemed it important that everyone had read and understood all of the steps involved. On page 1, there was a cautionary statement: "Do NOT

Figure 11.7. The interior of the new addition.

attempt to assemble an Ash Dome unit during periods of high or gusting winds." If you ever plan to assemble one of these, you will want to heed this warning.

Day1 : Dome Skirt

The first major section of the dome to be constructed was the dome skirt. This is a circular wall assembly that has wheels inserted into a rail system allowing the dome to continuously spin. The wall plate assembly consists of circular segments which, when fastened together, form a continuous circular base plate with the dome roller fixtures attached. This was also one of the challenges the architect had to solve—how to mount a circle onto an octagon. The solution was to circumscribe the circle around the octagon. This would then allow rain to roll off the circular dome and not cause any water pooling problems.

The carpenters mounted 1½ inch thick by 14 inch wide top plate boards around the top of the octagonal walls of the tower. There were two layers of top plate boards glued together in an overlapping manner, resulting in a 3-inch-thick bonded plate. To circularize this, a custom jig was constructed that was centered at the concrete pier. A router was attached to the end of the jig and was used to cut a perfect circle the same diameter as the dome base plate. When the routing was complete, the result was a 3-inch-thick top plate with a circular outside shape and octagonal inside shape that was firmly attached to the top octagonal walls of the tower.

The dome base plate was attached to the top plate with ½ inch diameter bolts. At the center of each of the octagonal faces and under the top plate, a 3-foot long 5 inch by 5 inch by ½ inch steel angle iron was connected to each of the vertical walls. The dome base plate was then connected with a vertical lag bolt through the angle iron. This design provided more than sufficient strength to connect the 2,800-pound circular dome to the octagonal tower walls with no fear of high winds blowing the dome away (Fig. 11.8).

The entire dome base plate had to be perfectly level so the dome would spin with ease. The dome roller shafts were installed around the base plate so the dome support rollers (or wheels) extended over the edge of the structure. A total of 28 wheels were installed around the circumference of the ring.

The dome track rails were installed next. These semi-circular segments were installed over the rollers and connected together to form a perfect circle that freely revolved. The dome skirt was then bolted to the dome track. The skirt was made of 14-gauge galvanized steel that sat on the outside edge of the dome track rail.

The last part of the dome skirt assembly was the installation of the azimuth drive gear rack. This was attached to the inside of the dome skirt. The square holes in this rack allowed the azimuth motor teeth to easily spin the dome.

Figure 11.8. The dome skirt will allow the dome to easily spin.

At the end of the first day, a shiny circular ring had been successfully installed atop the octagonal tower. Despite its size, the ring could be spun easily with just the push of a finger.

Day2 : Dome Panels

The second day of dome construction was by far the most memorable. This was the day the structure actually began to look like an observatory. This was the day the dome panels would be installed.

There were three of us working on this phase of the construction. We set up scaffolding inside the observatory room. Unfortunately, the stairs to gain access to the observatory had not yet been installed, so the only way to gain access to the second floor room was via an extension ladder. The pallet with the dome parts was in the corner of the yard approximately 30 feet from the base of the tower.

Tom, the foreman and master carpenter, and I were positioned atop the scaffolding. Joe, the youngest carpenter and most energetic, was given the task of bringing each dome panel from the corner of the backyard to the scaffolding on the second floor of the tower. The dome panels for a 16-foot Ash-Dome are a little over 8 feet in length, and each panel is numbered and must be installed in a very specific sequence.

Since this was early on in the construction phase, the windows and doors were not yet installed. Joe would pull a panel off the pallet, carry it to the base of the structure, and pass it through the window opening in the first floor of the tower. Then he would go around the tower to the sliding glass door opening in the new addition to enter the building. Once in the lower tower room, he would grab the panel, carry the panel up the ladder, and pass it to Tom and me, who were atop the scaffold.

Each panel sits flush on the outside trim angle of the dome skirt. Tom stood on the scaffolding in the center of the dome cylinder and held the top of the dome panel while I aligned the screw holes in the dome skirt assembly. While I was aligning the bolts and tightening the nuts, Joe would descend the ladder and return to the dome pallet for the next panel.

The bottom of each successive panel was entered into the top of the preceding panel's rib. The panel was slid downward through the interlocking rib joint until it came to rest in a position on the dome skirt assembly. Based on Ash Dome's suggestion, the lubricant we used to make the panels slide more easily together was Ivory liquid soap. Oil would stain the roof panels, while a clear liquid soap would wash away in time.

While Joe was bringing the next panel for installation, I would tighten each panel in place. Since the dome could not support itself early on, Tom had to hold the top of the panels to provide support. It wasn't until two thirds of the panels were in place that the dome could actually support itself.

The repetitive process of installing the 35 dome panels took the entire day. When we were installing the last panel, the sky turned dark and the winds started to increase in strength. Then it started to rain, accompanied by lightning and thunder.

There were some final nuts and bolts that required tightening before we could conclude for the day. All three of us were working in the open metal dome structure with rain coming down, wind blowing, and lightning in the area. It was a little scary as we wondered if the unfinished dome could withstand the gusting wind or, worse yet, if we were going to become human lightning rods. Fortunately, the dome, though still incomplete, held up rather nicely under the increased winds, and we were able to complete the day's work without incident.

At the conclusion of the work day, the construction crew and I gathered in the yard and admired our handiwork. We were in awe of what we had accomplished and were amazed by the sight of the glistening dome towering over the house.

We were also covered with Ivory liquid soap from the lubricating process. You would think three grown men working in the hot July sun all day would reek of sweat. On the contrary, we all smelled Ivory clean.

When Barb came home from work, she couldn't believe how much we accomplished that day and how different the project looked. The observatory dome was visible for the first time from the front yard. It was an impressive sight. Barb also complimented me on how nice and fresh I smelled.

Day3 : Rails

Installation of the rails, or shutter track system, was the activity for the third day. Each shutter track rail is made up of two quarter circles of fabricated track. These quarter circles are bolted together and then raised into position over the top of the dome. After being connected to the dome, the upper shutter door will ride these rails while opening and closing. But, since the dome contains ribs and is not perfectly smooth, the track does not hug the exterior of the dome. It rides over the tops of the ribs at non-uniform locations along the length of the track. Ash Dome created unique spacers known as trim angles that had to be individually bolted into place between the shutter track and the dome at the rib intersection positions. The tedious process of bolting over four dozen of these trim angles into place is what took most of the day.

The motor bar was also installed at the top of the dome. The electrical motor to control the opening and closing of the shutter door was also mounted on the motor bar at this time.

At the conclusion of the day, the dome had its shutter tracks and was ready to accept the installation of the upper and lower shutter doors.

Day4 : Shutter Doors (Upper and Lower)

Day 4 was devoted to the installation of the upper and lower aperture shutter doors. Before Ash-Dome shipped the 2,800-pound pallet with the 16-foot dome, I asked them what the heaviest piece was. Of the hundreds of pieces, the single heaviest piece was the upper shutter door, weighing in at 180 pounds. I shuddered at the thought of how we were going to raise a 180-pound door over 25 feet to the top of the tower and install it on the shutter track rail system.

Actually, Tom devised a very simple solution to this problem. He had a rope and a manual winch with a ratchet. The winch was attached at the motor bar installed the previous day at the top of the dome. A long rope was threaded through the winch and dropped to the ground where the upper shutter door was tied. On this day Eric, another one of the carpenters, was enlisted to assist with the dome assembly.

Eric and Joe were positioned on ladders along the side of the tower. Their role was to safely guide the shutter as it was raised up the side of the tower. Tom and I were in the observatory room and would pull on the rope to raise the shutter into position. The idea was to raise the shutter over the open aperture. When safely stowed and locked into position (held firm by the rope and winch), the rollers were inserted into the shutter tracks, the mounting fixtures were slid over the shaft of the rollers, and the fixtures were attached to the shutter. There are only six of these rollers required for a smooth operation. When all of the rollers were in place, the shutter section was slid upward so the shutter drive track was positioned over the upper drive gear unit. Once the track and gear teeth were properly meshed together, the shutter door would not move unless commanded to do so by the motor.

The lower shutter door, considerably smaller and lighter, was easily installed. A similar set up with the rope and winch was used to raise this smaller door to its correct position. Instead of rollers that rode the tracks, the lower shutter door was attached to the bottom of the dome by a hinge.

Day5 : Motors and Control Boxes

The end of the dome construction was definitely in sight as dawn came on the fifth day. The last items to install were the azimuth motor to rotate the dome and winch motor to open and close the lower shutter. The upper shutter door motor had been installed during an earlier step. Connecting all of these motors together were two control boxes that contained the operational logic. Ash Dome did a very nice job of ensuring that you could not close the upper shutter door before the lower shutter door was properly stowed. The control boxes were designed to only let you operate the dome in a safe manner. The remote controls for the dome operation also communicated with the control boxes.

Attached to the dome skirt were three power contactor bars. The main power was transferred to these contactor bars through floating power pads. Power to the control box for the upper aperture door motor and lower aperture door winch was obtained from the main power bus through the pads and onto the contactor bars. Then, no matter how the dome was last positioned or even while the dome is rotating, the aperture doors could be electrically operated because power continuously encircles the base of the dome.

Figure 11.9 shows how the winch motor (left) lowers and raises the lower shutter door while the control box (right) controls manual and remote operation of both lower and upper aperture doors.

Ash-Dome offered an option that would allow the shutter doors to be manually operated, but I did not select this capability. The reason why will be explained later.

At the end of the day, the aperture door weather seals were in place, the lubrication of all parts was completed, and the azimuth drive seals were installed. We came to the section in the installation manual that said

Construction of your Ash Dome is now complete. Congratulations!

I was elated at this point in time. The dome was done. It took 5 days, but it was weather tight, operational, and it looked good. Now it was time to put the finishing touches on the project to make it a truly first-class observatory.

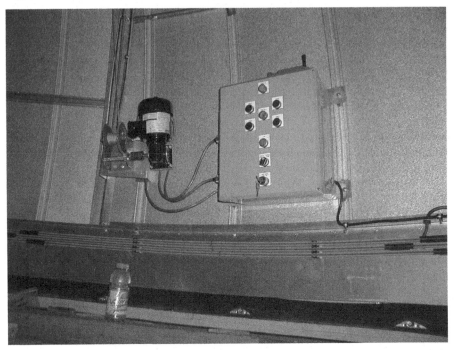

Figure 11.9. The winch motor.

Electricity

Electricity is woven into the fabric of our society, including modern astronomy. In an observatory, it is important to have a sufficient number of electrical receptacles for various types of equipment and accessories. It was time to work with the electrician, explaining my requirements. It was interesting to note that the architectural drawings did not include any electrical details. That could have been done, but I opted to meet with the electrician and explain exactly what I wanted. However, before this meeting, I spent many hours deciding what electrical features I wanted by making notes and diagrams.

Interior Lights/Dimmer Switches

Every astronomer knows that white light is bad for night vision, while red light will preserve night vision. Actually, instead of saying red light preserves night vision, it might be better to say that red light does not remove the night vision your eyes have attained. The lighting system in the interior of the observatory was therefore designed with this in mind. In order to gain access to the observatory, you enter through a door on the first floor and climb a set of stairs that hug the inside four walls of the tower. The stairwell contained four lights: two red lights and two white lights. Two separate switches would control each set of colored lights. The switches would be positioned in three different locations: at the bottom of the stairs outside the observatory entrance door; at the bottom of the stairs inside this door; and at the top of the stairs at the observatory landing. A dimmer switch would also be installed upstairs for the red light system.

A dimmer switch is a must for any observatory. Invariably, the brightness of the red lights should be controllable. I find it very useful to start the observing session with the red lights on full. This allows my eyes to begin their night vision adjustment. It doesn't take long when I begin to dial down the intensity of the red lights with the dimmer.

In the observatory, there is a window on each face of the octagon. The eight windows let in sufficient sunlight so electric lights would not be required during the day. An additional benefit the windows offer is the ability to cool the dome during the warmer months.

A light would be installed above each window: four red lights and four white lights alternating every other window. Because the walls were lower than standard height walls, I selected outdoor deck lights that were encased in a metal frame. This would help prevent damage to the light bulbs should anyone bump into the lower mounted fixtures. Again, two separate switches would be installed to control the different colored lights. The switches would be positioned in two locations: at the top of the stairs and at the base of the telescope. Thus, if you are observing and need to turn on any lights, you do not have to cross the room to accomplish this. A dimmer switch was installed for the red lights in the main observatory as well.

With this lighting design, it is very easy to turn on a light, either red or white, before ascending the stairs. Once in the observatory, another set of switches allows activating and dimming the set of red or white lights for the room.

Light Control

During the holiday season, Barb thought the observatory tower would look nice if every window contained an electric candle. I agreed with her and did not object to her design request. However, this meant I had to devise a way to easily turn off every lighted candle prior to the start of an observing session. Under every window in the observatory room, I placed an electrical receptacle. I requested the electrician to have the lower receptacle to always be "hot." The upper receptacle, however, would be controlled by a switch. By designing the receptacles this way, Barb could have her electric candles, which would turn on automatically when it got dark outside. Whenever I wanted to observe, I would simply turn off the switch to those receptacles to easily extinguish the electric candles.

Electric Telescoping Pier

While designing the wall height in the observatory room, I needed to arrive at a height that would allow me to see objects near the horizon comfortably. I also required the telescope to be at a comfortable viewing height for objects straight overhead. The telescoping Pier-Tech 2 offered a perfect solution to these design issues.

The Pier-Tech 2 is an electro-mechanical telescope pier. It was designed for the observer who wishes to raise and lower the pier by a push of a button. By pushing the button on the hand controller, the pier will raise an additional 20 inches in height. The precision vertical adjustment keeps the telescope polar aligned and level while keeping the target within the field of view. The Pier-Tech 2 lifting capacity is 215 pounds.

If you are interested in designing a first-class observatory, a pier similar to this is a must. Having one allows you to raise the telescope to easily view objects directly overhead without having to strain your back and be a contortionist. Plus, you can easily and comfortably look at objects near the horizon by positioning the telescope to the desired height.

We have also found this to be very convenient for visitors to our observatory. The telescope can easily be lowered for smaller children and raised up for adults, allowing comfortable viewing for everyone.

The Pier Octagonal Base

The concrete pier emerging from the observatory room floor presented an interesting aesthetic problem. When you enter the observatory room, the first thing that catches your eye is the immensity of the 16-foot diameter dome. All visitors to the dome are amazed at being in a room with a hemispherical roof that is 14½ feet high. Next, the eye is drawn to the center of the room where the telescope is located. The telescope, of course, is mounted on the Pier-Tech 2 (Fig. 11.10). The Pier-Tech 2 is bolted to the steel plate that is firmly affixed into the top of the concrete pier. One foot of the concrete pier is above the floor of the observatory. Seeing the exposed concrete just wasn't visually appealing. Therefore, I decided to build a pier base to hide the top portion of the concrete.

Being somewhat handy with woodworking, I wanted to build a base for the pier that would fit with the décor of the observatory room. I was originally going to build a simple cubical box to hide the top of the concrete pier. However, Barb suggested another alternative. She thought it would look nicer if the "box" was octagonal in shape to match the shape of the room.

My first inclination was that building an octagonal box was going to be rather tricky. However, after discussing the details of construction with a friend at work, the task turned out to be actually simpler than I had first thought. The octagonal box was constructed out of oak. Each edge was cut to 22 degrees 30 minutes. The edges were joined together using a biscuit joiner. The top of the box was composed of four pieces of oak joined together and cut in the shape of an octagon. The top edges were routed to soften the appearance of the box. In two faces of the box sides, I cut square holes. The first square hole housed a quad receptacle so I would have electrical power at the base of the telescope. The second square hole would be used for the light switches. I also have an Ethernet cable at the base of the telescope for communication with the control room, also known as the warm room, which is located on the first floor of the tower.

I made the box large enough to easily accommodate four coasters. I find this a convenient place to rest a steaming hot mug of cocoa in the winter or refreshing iced tea in the summer (or perhaps even a snifter of brandy year round).

Figure 11.10. The Pier-Tech 2 telescoping pier.

In the top of the box, I drilled four holes for the bolts of the Pier-Tech 2 to go through. With this design, the pier octagonal base box covered the top of the concrete pier but still did not make any contact with the telescope or concrete at any point, again achieving the requirement that no part of the structure touch the concrete pier. Finally, the box was stained and finished to match the window trim and railings.

Vibration Test

When Ed and his team were finished with construction, it was time to mount the telescope on the Pier-Tech 2. But before doing this, Barb and I decided to do the vibration test. We filled a glass with water and placed the glass on the top of the Pier-Tech 2. I then placed a flashlight next to the glass so the beam would shine through the glass and reflect off the bottom of the water. With everything set, Barb and I then began jumping up and down on the floor next to the pier. It was the strangest dance I ever did with my wife. However, we were both very delighted to see the water remain flat as a sheet of glass. We both knew the design of the building and pier column were isolated from each other, but this test proved the construction had indeed been done correctly. We had a pier completely isolated from the structure and extremely stable.

Finishing Touches

There are several other amenities that were added to provide that additional "touch of class." This section will discuss those final finishing touches.

Carpet

The floor of the observatory room was padded and carpeted with wall to wall indoor/outdoor carpet. There were several reasons for the carpet. First, when you are observing, the padded carpet is more comfortable to stand on for considerable periods of time. It also offers more warmth during the winter season. And finally, if ever I drop an eyepiece, instead of it shattering or becoming damaged, there is a greater chance it will bounce and survive unscathed. It is highly recommended for anyone designing their own observatory to consider adding carpet.

Computer and Telephone

I pre-wired the observatory for both computer and telephone. About the only wiring I did not bring up to the observatory was the television cable. And that was because the views through the telescope are more impressive, more incredible, and more soothing than anything available on TV.

Thermostat for Heating and Air Conditioning

The observatory room has its own thermostat to control both heat and air conditioning, but this was not done for the comfort of the observer. During observing sessions, you do not want to attempt to alter the temperature of the room, since the viewing would be severely degraded by heat current eddies. The temperature of the room should be the same as the outside temperature. However, during periods of excessive cold, heat, or humidity when we are not observing, it is desirable to have the ability to pump a bit of warm or cool air into the observatory for the benefit of the equipment.

Music

I really enjoy listening to music while observing, and I took this into account while designing my observatory. Two in-wall speakers were embedded between the studs during construction. The music system located in the new addition has a switch that allows me to pipe music just downstairs, just in the observatory, or in both rooms. I love to play planetarium music, which is the perfect musical accompaniment while observing the heavens.

Wireless Thermometer

In the first floor room of the tower, there is a thermometer that displays the temperature of the observatory room via a wireless thermometer. This enables you to get a good idea how to dress before you ascend the stairs. Of course, the beauty of an attached observatory is that it is very easy to come back downstairs and grab an extra sweater if it is cooler than expected.

Rolling Desk

Every telescope has accessories such as eyepieces, solar filters, camera attachments, etc. For my observatory, I purchased a seven-drawer office chest on wheels. This serves as a mobile platform that not only stores all of my telescope and camera accessories but also acts as a desk for my laptop when I connect it to the telescope. I simply roll the desk near the telescope, pull up my folding chair, and start the imaging process. At the conclusion of an imaging session, the desk is rolled back to its parking location.

Swiveling Floor Loungers

Another useful item to have in the observatory is a five-position swiveling floor lounger. We have a pair of these on the floor and have discovered they are extremely comfortable. The five position adjustment allows four different angles of recline or you can set it completely straight. These are great to relax on while viewing a small portion of the sky through the aperture opening or to use while reading a book on a weekend afternoon during the warmer months.

Control Room

The room below the observatory is called the control room or warm room. It is octagonal in shape and has a 2-foot diameter concrete telescope pier going from floor to ceiling in the center of the room. We decided to surround the concrete cylinder with sheet rock. This serves two purposes. First, the room looks much nicer with the concrete hidden and second, no one can accidentally bump into the pier to transfer vibrations to the telescope. This was the finishing touch to truly isolate the pier from the structure.

Dan, the architect, jokingly suggested that the concrete pier be wrapped with rope from floor to ceiling to make the world's largest cat scratching post. It would have been a novelty, and our cats would have loved it. However, if they decided to use the post while I was imaging, there would have been noticeable vibrations, so I laughingly rejected this suggestion.

Three of the walls in the control room are floor to ceiling oak book shelves. This allows ample room for storage of astronomical reference manuals, observer guides, almanacs and other space related books. The fourth wall contains a built-in desk with additional book shelves above it.

Someday I may decide to use the control room for remote viewing via the computer. Right now though, I much prefer being in the dome and observing directly through the telescope no matter what the temperature is. However, the control room does serve as a very nice home office.

Generator

I could have ordered the manual override controls from Ash Dome. However, since I didn't, the opening of the upper and lower shutters on the dome and the rotation of the dome is dependent on electricity. What would happen if the dome were open and there was a power failure? Or, more importantly, what if there were a major power blackout and the skies were black to the horizon because no one had lights? In either scenario, it would not be possible to close or open the dome, respectively because the power was out. We already had a generator, so the dome controls and telescope power were wired so they could be independently powered via that generator. I would simply have to start it up, flip a switch, and the dome would be fully operational.

Visitor's Log

We have a visitor's log book and request that every visitor to the observatory sign it. When visitors come down stairs after viewing, I ask them to sign, date, and if they wish provide comments on their experience. The most common comments are Wow!, Fantastic!, Awesome!, Cool!, Beautiful!, Amazing!, Incredible!, and my favorite "The only difference between men and boys is the price of their toys!"

First light for our observatory was October 28, 2005. My entry to the visitor's log was

"First Light"! Absolutely fantastic! Viewed Polaris, Mars (almost at opposition), Ring Nebula in Lyra, double star cluster in Perseus. Simply wonderful!

And this was followed by my wife's entry:

"What a beautiful evening! Our dream has come true."

Many thanks to my wife, Barb, for helping my dream come true.

Naming the Observatory

The observatory needed a name. This was not something to be taken lightly. Barb and I thought about this very carefully. We finally decided upon calling our observatory "Stargate 4173 at Grimaldi Tower." We chose this name for the following reasons:

Stargate

This observatory is a gateway to the stars. And I have been a science fiction fan for most of my life. Stargate was the name of a popular 1994 science fiction movie and subsequent TV series. Since science fiction was the major force that got me interested in astronomy, Stargate seemed apropos.

4173

The latitude and longitude of the location of the observatory is 41 degrees 36 minutes 49 seconds North, 73 degrees 40 minutes 16 seconds West. The degree portion of the coordinates, 41N 73W, was selected to uniquely identify our observatory.

Grimaldi Tower

Most people assume our tower was named after Francesco M. Grimaldi, who was an Italian physicist and astronomer, and author of a map of the Moon which was used by Riccioli as a basis for his nomenclature. However, that was not the reason the name Grimaldi was selected. Three weeks prior to the beginning of construction, one of our precious cats passed away, due to complications from feline HCM (hypertrophic cardio myopathy). His name was Grimaldi, and in his memory we named the tower after him.

As a personal side note, our other two cats are named Copernicus and Tycho. All three of our cats were named after craters on the Moon. So, in a sense, our cat Grimaldi was named after Francesco M. Grimaldi.

The finished observatory can be seen in Fig. 11.11.

Figure 11.11. Grimaldi observatory ready to observe the sun.

Personal Observations: Dome versus Roll-Off Roof

Here are some personal observations regarding domed observatories versus roll-off roof observatories. I do not have a roll-off roof, so comments relating to this style are my opinion only.

Advantages

Dome: A dome says "an astronomer lives here." A dome offers better protection from the wind. This is important for both the observer and camera equipment attempting to capture an image without worry of wind-induced movement. Since we live on a rather windy hill top, this influenced our decision to select the dome. There is also little to no dewing problems, as the dome helps protect everything contained inside. Also, the lower aperture door serves as a light shield and additional wind block.

Roll-Off Roof: You have complete access to the entire sky above. You can operate several instruments at the same time in the roll-off roof environment. Construction of a roll-off roof can be less expensive as compared to the domed observatory.

Disadvantages

Dome: You do not have complete access to the sky. You are viewing only a fairly small portion of the sky through a slit in the dome. This is not great if you want to view a meteor shower or follow the space station or shuttle. To see another portion of the sky, you must rotate the open aperture to the field of view of interest. Also, you are primarily limited to a single instrument located at the pier location.

Roll-Off Roof: You are more exposed to the wind and stray light. You need twice the amount of space because you need to roll the roof to another position for storage. There is a greater chance that rain will blow in along the sides as well as at the ends of the roof. The roof requires a detailed design analysis to prevent it from blowing off in both the open and closed positions.

Would I Have Done Anything Different?

This has been a common question from visitors to our observatory. Would I have done anything different if I could do it over again? The answer is simple—no, except to have built it sooner so I could have more time to enjoy it. There was a lot of thought and research that went into the planning and design of the Stargate 4173 at Grimaldi Tower Observatory. I spent about two years thinking about the design. If you want a first-class domed observatory, take the time to thoroughly think through all of the options available to you. By doing so, the end result will provide you with years of satisfaction.

Additional information and images of Stargate 4173 at Grimaldi Tower can be found at the Web site: http://www.stargate4173.com/

Resources

The following were very helpful to me during the time I was thinking about the design of my observatory.

[1] Patrick Moore (Ed.) (1996). *Small Astronomical Observatories*, Springer-Verlag London Limited.

[2] Patrick Moore (Ed.) (2002). *More Small Astronomical Observatories*, Springer-Verlag London Limited.

[3] Amateur Astronomical Observatories at http://obs.nineplanets.org/obs/obslist.html

CHAPTER TWELVE

The Spring Gulch Observatory, Jackson Hole, Wyoming

Brad Mead at 43°29′46″ North and 111°47′20″ West

For a little more than a century my family has raised cattle at the base of the Teton Mountains, in Jackson Hole, Wyoming, where the combination of high altitude and dark skies provide the dedicated deep-sky observer or dedicated imager with real first-class raw material.

And you do have to be dedicated. Nights from the end of November through the middle of March can be bitter; and while the sky is clear and dark and transparent, the clear skies and high altitude exact their toll on the thermometer.

Those dark nights—winter and summer—piqued my interest in astronomy. Ranchers spend a good deal of time out of doors, and, like the shepherds in the Arabian desert, learn something of the night sky simply because it is always there. The skies above the ranch are relatively free of light pollution, and we do a lot of our work after sundown. When we are calving in the spring of the year, we maintain a 24-hour watch on the cattle; mother cows are perversely more apt to have trouble with their calves between midnight and four in the morning. During the summer, we bale hay late into the night. And during the long, cold winter, when we feed the summer's hay, our afternoons are mostly free, affording an opportunity to plan and take a short nap. All of these activities take place in the meadow next to our house. Ranching lends itself to astronomy.

My interest in astronomy was also furthered and encouraged by my dad, an amateur astronomer himself, whose 3½-inch Questar instilled in me an appreciation for exquisite optics and excellence in design and construction, and whose accounts of cosmology expanded my horizons.

Figure 12.1. The Spring Gulch Observatory.

Figure 12.1 shows the glorious location of the observatory.

Over time, my appreciation for high-quality telescopes manifested itself in the ownership of two fine Celestron Schmidt–Cassegrains, an 8-inch and an 11-inch; three refractors—a Televue Oracle, a 92 mm Brandon, and a 5½-inch Astrophysics; and, of course, the Maksutov–Cassegrains that started it all—I've owned two, a 7-inch and a 3½-inch Questar. What was missing in all of them, though, was an efficient and easily accessible permanent mount.

My wife Kate and I assumed responsibility for management of the family ranch about 10 years ago, and we moved into the old ranch home where I grew up, an extended two-story structure surrounded by mature fir trees and nestled up against the edge of one of our hay meadows. The building needed remodeling, beginning with our bedroom, and I seized at the opportunity to include an observatory in the remodel.

Seizing on the idea and implementing it were two different things. I was helped in the implementation process by a fortuitous visit to Great Britain with Kate and our two sons. While we were traveling, I kept in touch via e-mail with a Gordon Rogers, who had been featured in *More Small Astronomical Observatories*, a book in my library from Patrick Moore's Practical Astronomy series. As we happened to be traveling nearby, Mr. Rogers graciously invited us to visit him and his wife, Margaret, in Long Crendon, Buckinghamshire, and take a look at his observatory. That visit was a true inspiration and the highlight of our trip. Gordon and Margaret started by taking us to dinner and then gave us a thorough tour of his observatory set up. Gordon's Crendon Observatory is located in a picturesque small town, with historical significance and tradition that far exceeds that in our comparatively new settlement in Wyoming. Gordon was able, however, to create an observatory second to none and do important work there. He and Margaret were generous and engaging hosts. We left their home with new friends, and on our return to the United States we were committed to follow their example.

Based on my visit with Gordon and some independent research, I had some parameters I knew were important in the construction of a convenient observatory. I did not want to have to leave the house in order to get to the observatory. From experience with a smaller roll-off roof observatory that was unconnected to the house, I knew that deep snow could be a disincentive to trekking out for an evening's observing. The roll off roof was permanent, but it wasn't efficient or easily accessible. Maintaining my view towards the horizon was important to me too. The ranch is bounded on the eastern and western horizons by buttes deposited 200,000 years ago by Buffalo glacial ice, and I lose about 10 degrees of sky to both the east and the west as an unavoidable consequence. I wanted as little avoidable interference with the view of the horizon as I could manage. The construction would have to either remove or accommodate the mature trees around the house and, as much as practical, the roofline of the house itself.

I also had in mind a dome. A roll-off roof is wonderful. It provides a panorama of the sky and connects the observer in a particular way with the context of his work. But in a climate with bitter cold and frequent heavy snowfall, a dome often makes a night's work more possible—or certainly more palatable. I had used both roll-off roof and dome observatories and enjoyed both types. I concluded that I would like a dome, but would also like a platform connected with it that would afford me the opportunity to get outside the dome and enjoy a whole-sky view when the weather was nice.

Beyond these specifications, I knew that I wanted a separate control room, heated and comfortable. I wanted elbowroom around the telescope. And I wanted plenty of space for ready access to the atlases, catalogs, texts, and other equipment that form part of the advantage of a permanent installation.

With these criteria, the first and most important step in the construction was designing a structure that was both practical and aesthetically pleasing. After a careful search, we ended up deciding to work with an outstanding local architect,

Stephen Dynia, and asked him for help with both the bedroom and the observatory. Stephen quickly concluded that the observatory and bedroom should match the rest of the house in subtle design details, but in the main be unapologetically and dramatically different but aesthetically pleasing enough to pass muster with the local planning authorities. Stephen and his staff then had the job of selling this somewhat unusual structure to the County Planning Office.

To put Stephen's difficult job in context, Jackson Hole is a community that prides itself on maintaining a sense of its historical character. It is a place where old buildings and open space are revered. Local land use laws emphasize community character, the Western tradition of cattle ranches, and the use of building materials that are subtle and blend with the landscape. Local planners take their charge very seriously, and we have many neighbors on the hillsides east and west of the ranch that bought expensive lots and built costly homes so they could enjoy the view across the ranch and up toward the Tetons, about 10 miles north. In this context, we wondered what sort of reception our proposal to build a modern wing on the end of an old ranch house would have, particularly when the wing included a 10-foot aluminum Ash Dome. We decided to think positive, and began the design phase of the observatory, hoping our plans would not meet too much resistance.

While Stephen and his crew shepherded the house and observatory plans through the planning office, I began to shop for a telescope. At the time, Kate and I were living in an upstairs guest bedroom, and the entire north end of the house was demolished. Given the magnitude and expense of the project, I felt like we were in for a penny, in for a pound. I was committed to an instrument with first-class optics and uncompromising mechanical construction.

I finally settled on a 16-inch Ritchey-Chrétien built by Brad Ehrhorn at Ritchey Chrétien Optical Systems (RCOS) in Flagstaff, Arizona. This f8.4 telescope is guaranteed 1/25 wave RMS or better, incorporates active cooling, and is housed in a carbon fiber tube. The carbon fiber tube changes dimension very little when the temperature goes up and down, making night-to-night focusing easier. The delivery estimate for the telescope was four to seven months, and I wanted to be able to use the observatory as soon as possible after construction was finished. So I ordered it. Having paid the deposit on the telescope, we were committed!

While the building design and permitting process proceeded, I worried about making sure I had a clear view of the Southern skies. Jackson Hole is located at more than 43 degrees North latitude, so it's important to be able to turn the telescope to the south. I also, as mentioned before, wanted the best view possible to the east and west within the constraints imposed by the buttes in either direction. I decided I absolutely could not remove any of the old growth fir trees planted by my ancestors in the yard, so the only alternative was to raise the observatory high enough to see over them. The question was, how high was high enough?

To find out, I took the somewhat desperate step of wedging a large stepladder into the bucket of a back-hoe, perching myself on the stepladder, and having Kate raise me as high as the machine's bucket would go. From this irresponsible vantage point, I sighted the horizon through a paper towel tube and—raising the bucket of the back-hoe up and down -came to the conclusion that the bottom of the dome's aperture needed to be 23 feet off the ground if the trees weren't to interfere.

With Stephen's efforts and expertise as an architect, we finally got a permit to build this unusual structure. The observatory's dimensions complied with the local height restrictions and, somewhat surprisingly, the issue of compatibility with local traditional architecture did not come up. Perhaps the planner looking at the plans was an astronomy enthusiast!

Construction of the bedroom and observatory began in October 2003 (Figs. 12.2 and 12.3 leave no doubt that this was a winter operation!). Bill MacLeod, of MacLeod Construction, was the contractor and did yeoman's work with a challenging project, much different than the custom log homes he usually builds. As expected, construction went over time and over budget; we knew this would happen because of the specialized nature of the building and construction techniques that had not been used locally before. The first—and principal—challenge was construction of a tall pier adequate to provide stability at the height of the telescope and one that would be isolated from the inevitable vibrations of the house and observatory building itself. The engineering solution was concrete—lots of concrete, almost 54,000 pounds of it!

First, a concrete pad was poured on a carefully prepared base. The pier would be located in a part of the hay meadow and thus in topsoil. The topsoil was removed down to the level of rock and gravel and the concrete then poured on the solid underpinnings. Ultimately, the pad grew in extent to about 10 cubic yards, or about the size of a medium-sized automobile.

Affixed to the concrete pad were oversized bolts. These bolts were used to attach the pier itself to the pad. The pier consists of a 20-foot length of corrugated steel culvert, 16 inches in diameter, rebar-enforced, and filled with concrete. Together, the pier and

Figure 12.2. Early construction.

Figure 12.3. All weather builders.

pad are massive—massive enough, in fact, that the process of polar alignment had to be repeated again and again over a period of months as the concrete settled into the gravel and bedrock, tipping the long arm of the pier ever so slightly. After 6 months or so, the mass stabilized and has not moved since, even—so far as I can tell—during our frequent but mild earthquakes.

After the pad and pier were constructed, relatively normal construction proceeded on the bedroom itself while we awaited delivery of the dome from Ash Domes.

In March 2004, the dome arrived. To the credits of the architect and contractor, the truck carrying it showed up only about a week before we were ready to install it. Ash Dome will provide a service of on-site installation and connection of the dome gores, but Mr. MacLeod felt he could do the job, and with the excellent instructions provided by Ash Dome was able to have the entire dome up and rotating within a week. We had a moment or two of anxious nail biting as the dome's shutter was installed; the shutter included an optically superior pane of glass 36′ by 36′ that could easily have broken if dropped or mishandled from 25 feet above the ground.

Once the dome was finished, I started spending some very enjoyable and exciting hours in the observatory planning the interior arrangement of books, computers, cameras, and other tools and equipment necessary to a stand-alone convenient operation. From the photographs, you can see that the dome sits on the east end of a rectangular elevated structure. The west end is a control room, which is separated

Figure 12.4. The dome interior.

from the dome by an airtight door and contains bookshelves and the computer that operates the telescope, the mount, and the imaging camera (Fig. 12.4). This arrangement permits imaging from the warmth of the control room but also permits quick access to the telescope if necessary. The proximity of the control room also allows me to use the radio-controlled dome remote to rotate the dome and raise or lower the shutter without going in the dome itself. Avoiding the dome has both practical and ergonomic benefit. Entering the dome from a heated control room creates an image-destroying vortex of warm air escaping from the shutter; it also, in the winter in Jackson Hole, dilutes my enthusiasm for astronomy. We have many nights most winters where the temperature falls below minus 40 degrees Celsius. Those are, typically, nights of clear, steady, and dark skies, but they are also uncomfortable nights.

After selecting the telescope and designing the observatory, I ordered a Paramount ME to drive the telescope. This mount was, at the time, difficult to get, and I ended up on a 6-month waiting list. The mount was worth the wait. The machining is reminiscent of that on a Questar. Its performance, in extremely low temperatures, has been phenomenal. Coupled with Software Bisque's *Sky* program, the dome and the telescope have a foundation worthy of their potential.

The sole issue with the Ash Dome—at least at the time I ordered it—was an inability to link the dome's rotation with the Paramount ME's movement in right ascension. As long as I start both the telescope and the dome's shutter in the proper orientation, this does not present a problem. The dome's shutter is 4 feet wide, permitting relatively long exposures while the telescope moves in right ascension without the need to rotate the dome.

To ensure proper orientation of the dome shutter and the telescope without entering the dome, I adopted a simple solution: I point the telescope from the control room, and then, using an inexpensive web cam mounted at the back of the telescope, take a real-time look at the control room's computer to make sure that the dome shutter is properly positioned. This approach may seem somewhat archaic and blunt, and would not support all-night automated imaging. It is, though, a functional limitation of the observatory for two reasons: First, at the time I purchased the Ash Dome, there was the difficulty coordinating the telescope's movement in right ascension with the dome's rotation. Second, although I have friends computer-savvy enough to install encoders and micro switches that would permit such coordination, to my dismay, rotation of the dome MOVES THE TELESCOPE. This movement is not discernable to the naked eye—it is not even discernable at magnifications up to 100 diameters, but it does exist and would be a real detriment while imaging. The source of the movemenvt is low-frequency vibration transmitted by the dome's rotation down through the framework of the house and into the ground. From thence, the massive block of concrete is affected, and the long arm to the top of the pier magnifies that effect. If there is a lesson in this building experience it is to over-design stability and then, once over-designed, double the over-design.

So, how does the observatory work?

Very well! A telescope—any telescope—is a thing of rare consequence and transcendent beauty. A large telescope like this, with superior optics, a great mount, and a permanent and convenient shelter is an unbelievable luxury and a real opportunity: My observatory inspires me to active investigation of the night sky.

I use the wonderful Internet-available Clear Sky Alarm Clock (www.cleardarksky.com) to alert me to the potential for a night of good seeing. Whether I am putting up hay, calving, or just feeding the cattle, I can take advantage of the convenience of the observatory to participate in the changing tableau of the night sky.

Typically, this begins by opening the dome's shutter sometime before dusk and turning on RCOS's Telescope Command Center. My telescope has thermocouples and fans on both the primary and secondary mirrors. I set the Telescope Command Center for the ambient outside temperature and the fans automatically begin to cool the mirrors. From experience, I have found that by turning the shutter to the northeast—180 degrees away from the prevailing wind—the dome cools down more rapidly and more quickly approaches the local outside air temperature, probably a consequence of the Bernoulli effect and the low-pressure area created on the upper down-wind side of the dome.

After opening the dome and setting the Telescope Command Center to cool the primary and secondary mirrors, I leave the observatory, making sure the webcam is sending a signal, and close the door between the dome and the control room.

I return to the control room after dinner, boot up the computer, and get my star charts and observing plan out. Using *The Sky* software, I point the telescope and then use the webcam to rotate the dome so that the aperture of the shutter is placed to give maximum movement in right ascension without the need for further movement of the dome. Images are stored and later processed on the control room computer. I usually spend a relatively small part of the night imaging, then bundle up, get in the dome, and observe visually using nothing more than a red flashlight and a star atlas. The finder scope on the Ritchey Chrétien is a 5½-inch AstroPhysics refractor, and on many nights of poorer seeing, I spend my time using the mount's joystick and happily star-hopping with the smaller refractor and my bedraggled copy of the *Millennium Star Atlas*.

Figure 12.5. The observatory wing.

But, regardless of the excellence of the telescope or the sophistication of the mount and its shelter, some of the most memorable nights are when I abandon the telescopes altogether, orient the shutter of the dome to the west, get the stepladder, and climb out of the shutter and onto the deck and look—just look—at the night sky. The wing incorporating the dome is shown in Fig. 12.5.

CHAPTER THIRTEEN

Public Reaction

Having built and equipped the observatory it was, of course, my pride and joy, but I did not expect it to generate much interest with the general public: I was wrong.

After completion the first parcel delivery I had was from Omega and the Rastafarian driver was hooked. I had a very enjoyable 10 minutes showing him around. The next day the chimney sweep and his son arrived, and it so happens that they are space nuts, so there is another tour. E-mails started to arrive from people wishing to have a look. One was "out of the blue" from a Polish man, Stefan Zietara, who runs an astronomy column for the estate where he lives and has become a firm friend. My surgeon brought his young children. I have always had keen sight, but young eyes see so much more than those that have been around a while. Jupiter is the target that really brings this home to you. Many parents ask to bring their children, but in a few cases I think the youngsters (and parents, too, maybe) are more interested in the indoor carp pool and the fully landscaped model railway in the attic! (Yes, the pool and the railway are over-the-top as well).

Then there was the occasion that I had been waiting for: the first "new" lady to be escorted around the dome. She just happened to be a gorgeous red-headed Dutch actress. Explaining that her man must stay below I ascended first. When she arrived in the dome and laid eyes on the equipment I wrote down on a display board her first words "Oh my God." The next step was to close the flap over the stairs so that she was totally incarcerated in the dome. I had no preconceived idea of the routine I would adopt, but it developed naturally. Having opened the dome and found Jupiter I went onto red light. Having adjusted the steps to suit her I held her hand, dousing the light while she took her fill of the planet. Then, red light on, reposition the steps, having found Saturn, lights out again, and await the so satisfying cry of amazement when a human being for the first time sets eyes on that magnificent sight in real time. The man is then permitted into the dome and allowed a cursory look at the planets and other prominent targets.

Comments from a few of the very many ladies that have been on the receiving end of this treatment include

"Oh my God!" (frequently)

"Oh, it's like the dentist!"

"It's sensational!"

"It's Big and Black!"

"Wow—What exactly do you do with it?"

"Where do you look?"

This is the exciting bit

"It's huge, Gordon!"

"Wow—that is quite a bit of equipment

"My God, gee wiz!"

"Magnifique!" (from a Parisienne)

"It's bigger than I expected."

Each year I entertain a party of students from Oxford University. They come from around the world and are aged around 18–20 years. Some of the questions demonstrate a sharp mind, and I have had challenging and enjoyable conversations with a number of them.

As I am listed as a speaker for schools by the Royal Astronomical Society I do quite frequently put a Powerpoint presentation in front of youngsters. My wife has a life long friend, Rosemary, who taught at a Roman Catholic school, and she asked me to give my first ever talk to her class of 11-year-old boys and girls. She had slated in one hour but said "You will know in 5 minutes if you are going to make it!" As I walked into the classroom I knew I was onto a winner; the whole of one wall was covered with drawings of Roman pots. How boring compared with what I could talk about! Concerned about getting to the children's level I had taken with me a plastic bag containing a large partially inflated alien figure. I explained that I had run out of breath, and he was soon tight as a drum in the seat next to me. We then went through a naming ceremony. Many suggestions were discarded, and it had to be Rosemary's clever daughter Susie who came up with Ziggy, which was unanimously adopted. They loved it. There was a forest of arms with questions from all quarters. One of the deepest was from a girl who asked what I thought about God. This got the undivided attention of Margaret and Rosemary, who visibly relaxed when they heard me say that looking through my telescope I felt very close to God. Mothers had to wait for their progeny while the queue of questioners slowly abated. I find that every class has one or two bright sparks, and this one was no exception. David was a boy that would go places. He asked if I was famous: I assured him I was not, but having given him tasks to perform in the sky and seen his talent, maybe I will be famous one day when he says "Mr. Rogers put me on this path."

I give a yearly talk to my local school, and some of the children come to see the observatory. The outcome of the first talk was thirty thank you letters. Half had been written out of duty, but I was delighted to realize that the remainder had come from interested children. Their thoughts were from all manner of directions and certainly made me think. Enormous effort had gone into some, and in Figs. 13.1 and 13.2, two are reproduced.

Figure 13.1 shows a lady astronaut with rockets at her feet, her radio, her umbilical cord, and, of course, her bare midriff as dictated by fashion.

Figure 13.1. A lady astronaut.

I recommend that you look closely at Fig. 13.2.

Leo is 8 years old, and his mind races around the universe. His imaginary names are so good, I asked him if they were from computer games. "No, they just came to me." This picture is even better in color.

In every school talk the subject of black holes arises, and the children are fascinated by their mystery (as are grownups). I am always concerned to assuage any fears that the children feel. Having taken a vote on whether black holes are good or bad, always

Figure 13.2. A boy's mind charging round the universe.

with a pessimistic result, I announce that I think they are good, with the analogy that they are like the hub of a bicycle wheel holding everything together and forming a nucleus around which matter can rotate.

I was getting a dialog going with a class of 8-year-old children, and the questions were slow to come. I asked if they could guess how scientists knew the precise date when Messier 1was first seen from Earth. Two hands went up, and the first boy asked, "As you have told us about the expansion could you work back from this?" Brilliant, but you could not arrive at an exact date. The teacher seemed nervous about the imperious hand held aloft by Thomas, but as there was no other she selected him.

"Mr. Rogers, it is cheeky."
"Well go on then."
"Were you around at the time?"

At the Congregational Church we had listened to the organ and Bible and poetry readings and sung hymns before my time came. Conscious that time was short, I decided to skip some pictures; this was accepted but for my most rapid of attempts to flip a slide of a lovely fashion model with a see-through top. There was a clamor for me to go back, expose the picture, and give them the full story. Since it concerns astronomy I suppose it is legal to recite it here. I had an e-mail from RDF television. They make a television program for the UK's Channel 4 called "Faking It," where they take an innocent and turn him into a professional something or other. She wanted to turn an astronomer into a London fashion photographer. "There you go, you have hit the jackpot first time. When can I start?" She quickly indicated that I was too old, to which I responded that young astronomers do not exist. I mentioned Nick Szymanek, with his lengthy blond hair, to her but, though he looks every inch the part, he was rejected, too. They eventually gave the job to a radiographer who should have been doing more important things. As an aside, my wife's friend Rosemary, by that time, a theatrical agent, by chance acted as a judge when they were creating a magician. Now Rosemary never gets it wrong, but she did this time, even though her husband is a Magic Circle man.

The Women's Institute, Senior Division, was interesting. After we sang Jerusalem, I explained that as a young man I thought the man would lead and the lady would follow: I got that wrong. Then it dawned on me that if a woman wanted to do something you could delay it, but it would not go away, and finally, if you obeyed orders life became a panacea. On leaving, one of the younger ones told me how much she had enjoyed my talk and asked where could she get a man like me!

I have been privileged to give talks at the National Space Center in Leicester (where there are all manner of hands-on activities for children) and in the William Penney Theater at the Atomic Weapons establishment at Aldermaston. On one occasion, I followed an eminent American who lectured about "The Face" on Mars. He had reported to the Senate and, together with his mathematical hypothesis, made an extremely good case for the shape being manufactured. Of course, we now know otherwise following better imaging from different perspectives.

My local Astronomy Society was giving an open evening at Waddesdon Manor, a magnificent Rothschild chateau owned by the National Trust and not far from Long Crendon. I arrived to find the estate closed because of the fear of falling trees in the extreme weather. There were three individuals at the gate, so I invited them to follow me home, where they got to look at the treasures of the night through 16″ of aperture instead of 4 inches, and the seeing was surprisingly good as the storm abated.

CHAPTER FOURTEEN

A Few Last Thoughts

I remember my first Astronomy Society meeting at the famed Rutherford Appleton campus at Oxford, where they do all manner of exciting things and you can actually talk to people who make components for space probes. I arrived in my best bib and tucker and stood out like a sore thumb. Margaret always preaches restraint in dress, but these guys were in vivid reds and blues and greens. Hair was worn very long, and I wondered who these guys were. When it came to question time I found out: a bunch of brilliant guys who could hold their own with the professor.

Not long after this there was an open day at Culham, headquarters of the UK Atomic Energy Authority, where they are researching nuclear fusion. Picking up a bunch of mathematical papers I turned to the chap sitting next to me and asked if all the jargon made any sense to him. He replied that it did, as he had just retired as Director of Electrical Engineering!

Oxford University advertised a 1-day astronomy course with some top speakers. I chose to sit next to an interesting-looking man of foreign appearance. I later learned his name was Jose Ramon Lopez-Potillo. In Mexico, he had the precisely the same set up as me, with both dome and telescope. Since he was a man of style, I judged that he would have an attractive wife and suggested dinner, to which he agreed. Indeed, Mantina turned out to be gorgeous. On inquiring how she met Jose, she said it was at the palace. Asking what she was doing at the palace she said "Oh, but Jose is the President's son." There you go; you get all sorts in astronomy.

In London, I got to talk with the astronaut Jeff Hoffman, who had flown on the December 1993 *Hubble* servicing mission. At one point In EVA he had to remove nuts not designed for astronaut gloves, being aware that one tiny lost nut could jam *Endeavor's* doors.

I had occasion to send a package to renowned amateur astronomer Steve Mandel but omitted to include a "sent by" label. It was at the time when the anthrax postal scare was in full swing. Steve debated with himself and eventually decided that the

Hidden Valley Observatory was probably an unlikely target. We subsequently met up at the Cambridge Institute of Astronomy, where he gave an inspiring talk.

Around 1996, at an astronomy shop in London, I fell into conversation with a fellow browser. What did he do? "Look for supernovae."

"Found any yet?"
"No."

He turned out to be Mark Armstrong, who is now heading towards one hundred supernovae finds, not to mention the asteroids.

At one of the Astronomical Society meetings I got to talk with Colin Henshaw, who was in Zimbabwe at the time of SN1987A. I was intrigued with his account of what happened and asked him to set it down in writing. It reads:

"I first started by observing both Magellanic clouds as seen by the naked eye and binoculars, and the Web Society published my results. On February 24, 1987, I left my cottage to make routine observations of variable stars, when I noticed that the 30 Doradus Nebula appeared rather conspicuous to the naked eye. I observed the variables I had planned to look at, then remembered the LMC and decided to check 30 Doradus. I noticed a 4th magnitude star South preceding the nebula that I could not recall having seen before. I returned to my cottage and checked my atlases, and could not confirm it, reaffirming my suspicion that I had discovered a supernova. I drew up a sequence, went back, and observed it, making it about 4 m.6. I returned to my cottage and phoned Richard Fleet, who had telex links at his office in Harare, and hopefully he could communicate a message to the outside world. Unfortunately it was raining in Harare, and Richard had turned in for the night. His mother took the call, but as it was raining she did not go outside to his cottage and tell him. By the following morning she had forgotten about my call, and maybe, Richard later pointed out, if I had been a little more pushy about the importance of what I had seen, she would have gone out. I went back to my observing site and made three more estimates of the object, each one brighter than the previous. I had obviously caught it in the rise, and I thought I could expect it to reach magnitude 1. Light curves published later often show a vertical row of four points representing the observations I made that night. I also ran off some photographs. Two days later, Richard was listening to the BBC World Service when it made an announcement about a supernova in the Large Magellanic Cloud. Richards's mother then piped up, "Oh, Colin phoned on Tuesday night—something about a supernova in the Large Magellanic Cloud!"

The next morning Richard phoned me at school, telling me it was confirmed, and that other discoveries had been made in Chile and New Zealand, but it was still not known whether I was first or last. It turned out to be the latter. It was clear again that night, and to my amazement I found it had faded. I made a color estimate and noticed it was white.

When I got back to Zimbabwe two months later, the supernova had brightened to 2 m.6, and turned deep red. Then it began to decline, and I lost it in January 1989."

In the Pacific Ocean, off Costa Rica, I had established that the Swedish Captain, Geir-Arne Thue-Nilsen, was actually an astronomer. We arranged that as the Moon would have gone by 11.00 p.m. we would together give an astronomy talk. He would also do what no other captain would do and douse the nonessential lights of the ship. We got on deck 10 minutes beforehand to a glorious pitch black sky sprinkled with jewels. At 11.00, people started to arrive and, needless to say, so did the clouds—

standard procedure. As I write, Comet McNaught is low in the west following the sun down. I had hoped to image it, but there is just one slender bank of cloud in the whole sky, on the western horizon, of course!

We were due to visit Sydney on vacation, and I had to attempt to see the Anglo-Australian Telescope about 200 miles distant. A senior operator there, Steve Lee, was most accommodating and promised to show me around. Early on the arranged day I arrived at the shed at the back of Sydney airport. I was greeted by a man, in epaulets, with the question: "Coonabarabran?"

"Yes please."
"We have a problem."
"What sort of problem?"

"The airstrip at Coonabarabran is closed because, in taking off, a plane hit a pig, and there is now a plane with no undercarriage sitting on a dead pig."

On phoning Steve Lee the response was a very sanguine "Oh, they hit another pig." Trip cancelled. I could not expect Margaret to miss *Madame Butterfly* at the Opera House that night.

Here is a story of another disaster leading to stimulation. I had arranged a birthday treat for Margaret at the Plaza Hotel in New York, and from here we would board a ship for an East Coast colonial tour. There was a storm in London, and all flights were cancelled, so we stayed that night at the Heathrow Crown Plaza, which was rather a far cry form the New York version. Queuing for several hours in seeking to re-book, those in our part of the line became well acquainted as Margaret kept us supplied with sandwiches and wine. The guy behind me was going to Saudi Arabia to insure oil wells and the chap behind him was trying to get to Seattle, where he was involved with Boeing's AWACS side. The fellow in front worked for British Petroleum and was on his way to Anchorage. He told a dreadful story about a paint spillage costing the company $30 million.

Eventually we did arrive in New York with very little time before the ship was due to sail and, of course, without any bags. Like a whirlwind Margaret did Saks from top floor to basement in 20 minutes. We rushed to the ship, last passengers to arrive, but there was a delay because the new captain had been held up and, in addition, the ship spilled oil into the harbor and had to report this. There was a delay of 2 hours while the environmental people argued the cost of this. Eventually we left New York at night, which made the delay worthwhile. The weather was bad and the ship bounced about. The ship had a bug and Margaret caught it. On day 3, we got to Philadelphia. It was still pouring, the streets were awash, and even the Liberty Bell was cracked. Still no bags, and British Airways did not have any idea where they were. A lady on a ship with no clothes is a disaster, and I agreed that if nothing arrived by the fifth day at Baltimore we would go home. For the formal night they rigged me out with a dinner suit in which other passengers mistook me for an undertaker.

By the fifth day our fortune changed dramatically. The sun shone, the bugs went away, the bags arrived, and I met a man called Dick Underwood. Dick ran NASA photography during the Mercury, Apollo, and some of the shuttle missions. He is one of very few people who have first hand knowledge of all the *Apollo* astronauts since he instructed them all in photography and debriefed them on their return. He stood over the developing tray and watched as the photograph seen by more human beings than any other came into view: the picture of Earth as an orb. Margaret reckoned

Figure 14.1. Richard Underwood at Cape Canaveral with the author.

that there were now three of us on holiday together, because I was intrigued with the inside story from a man who was there and took every opportunity to meet with him. We did the tour of Cape Canaveral together, where Dick used to commute to from Houston at times of launch (Fig. 14.1).

Dick's family background makes for interesting reading. His grandfather was killed when there was an explosion at the torpedo factory at Alexandria (Washington). His father was meteorological officer on the USS Hornet on the Doolittle mission to bomb Tokyo following Pearl Harbor and was responsible for the decision to launch the attack early because of his forecast of deteriorating conditions over the target. At the atomic bomb test at Bikini atoll cameras were banned, but Dick made his own and used a film sent to him by his mother secreted in a chocolate bar.

He married Rosa, daughter of the president of Honduras. He has visited every state and county in the United States and every county in the United Kingdom. Here are a few of the events he told me about:

1. *Apollo 8* had just rounded the Moon, and Bill Anders asked to speak to Dick. "What if we exposed the film as ASA 60?"
   "But you know it is ASA 10,000."
   "Well, we were tired, we were excited, and I gone done it."
   "Thanks for letting me know."

Dick researched the treatment of over exposed film and found a paper by Dr. C.E.K Mees. Dr. Mees was employed by an English company, Wrattan and Wainright Ltd. George Eastman of Kodak heard of Dr. Mees's talents and offered him a job, which he declined. So Eastman bought the company and appointed Mees as chief of research in Rochester, New York. From the papers written by the long-deceased Mees, Dick established how to develop the grossly overexposed film, resulting in some decent pictures from humankind's first pass behind the Moon.

2. Dick was at home in Houston one night when there was a tap on his window from one of the Capcoms who lived next door. "There is trouble at the office and we must go in." Dick says that in Control there is always an "odds of success" summary. When John Glenn first flew, the odds were just 6–10 in favor of success. As he entered the room the success rating for the *Apollo 13* mission was 1 in 999,999! You might remember that there was a carbon monoxide problem and a quandary about how to join a rectangular section with a circular one. It was Dick's idea to take one of his photographic plates, roll it into a tube, and grey tape each end to the respective orifices.

3. The controller, Gene Kranz, instructed that all personnel should remove their badges; rank was not going to hamper the rescue of his astronauts. Two young engineers, fresh from college, took a tea break and pondered how the stricken craft could be returned. On a paper napkin they came up with the idea of a never-before-tried double burn for the quickest return trajectory: Kranz went for it.

4. It was important to try to get pictures of the damaged service module, but it would be spinning wildly. After 23 hours in the mock-up, Dick arrived at a way to do this. He asked Jim Lovell to take one of the graduated photographic plates and drill a hole at particular coordinates. Then he asked him to clear the frost from the window with a medical wipe and, holding the plate in a particular alignment, to mark the window through the pinhole and allow it to freeze over again. At the appropriate time Lovell was to again clear the window, being careful to retain the marked spot and to hold the camera touching the spot and with a corner of the camera on a designated rivet. The camera was a telephoto one with a 36-shot reel. Thirty seconds after separation Lovell was to use all 36 shots. One of these had a picture that captured what had transpired.

5. On *Apollo 14* Al Shepherd was seen to make a motion like a golf swing. Asked what he had done he said he had played the first golf shot on the Moon. In answer to the question about how far the ball went, he said it was out of sight. Not so: I have a photograph of the ball 62 feet away, and the club is at a slightly lesser distance, an experience which many club golfers will have known.

6. Dick has, in a bank vault, a St. Christopher's medal that has been certified by one of the crew of each Moon lander to have accompanied him on the journey. I have been fortunate to receive from Dick a certificate incorporating a snip of film that went to Tranquility Base.

7. Four watt Trinitron television was a NASA development for the *Apollo* program. NASA offered it to American TV companies, who all turned it down; not so Sony, who developed it to the full.

8. From the vantage point of space, Dick watched as his pictures showed changes on Earth. He saw the Aral Sea in Kazakhstan, fourth largest in the world, shrink to half its size. He saw Lake Chad in Nigeria almost disappear over the years. In Zimbabwe, the rain-bearing cumulus nimbus gave way to dry cirrus. The North Polar ice cap was suffering a yearly retreat.

Conclusions

Having an astronomical dome, although taking you out of circulation while operating it, does give a social boost. When you get to be known as a "star" man, this topic of conversation is inevitably raised.

"Anything happening up there?" is a stock question. Any answer must be the understatement of the millennium, but invariably there is some recent, probably Solar System, event that you can relate. People also try out snippets they have read about in the papers, and you can see the little smile of triumph when they alight on something that has escaped me.

Now that I have the big comfortable permanent set up I would not want to go back to mobile imaging, but maybe I am too spoiled and losing out to the all weather guys. They charge off to remote sites and face whatever the weather throws at them. Tents are blown down, they get soaked, and they can go nights without a sky. However, when the clouds do clear and everyone gets going with their individual equipment, a camaraderie develops in sharing deep-sky treasures; unusually, for this pastime, there is an expert on hand to whom you can immediately refer for any little glitch that comes your way. The ultimate star party find must have been Steve Lee, of the Anglo Australian Telescope, finding his comet at a star party in Australia. He first thought it was a globular cluster and then realized none existed at the location in question.

Though I am a social animal, I would not be without my observatory. Still, I have to make a serious aside. If your sole objective is to take high-quality images of deep space, then forget building an observatory in England or other cloud or light polluted lands. Find yourself a partner or perhaps a commercial operator and arrange time on a robotic setup in an elevated clear sky site. Personally, I would choose some-where in the southern Hemisphere, where the delights of the Magellanic Clouds and the Southern Milky Way ride high in the sky. I would also select a location where dawn arrives at around midnight GMT to give the opportunity to turn in at a decent hour—What bliss! More and more amateurs are taking this route, and though they are producing fewer images, what does arrive is of much higher quality. Some sit in a blaze of light in Los Angeles and turn out mouth-watering pictures from their site in New Mexico.

Having said all this, there is something special about looking at, or imaging, an astronomical object from your own home. A couple of months ago I had gorgeous views of Comet MacNaught. Saturn, real time, is always breathtaking on a good night. Capturing light that has travelled 9 billion years from a quasar before striking your mirrors has to be evocative.

CHAPTER FIFTEEN

Resources

Astrophotographers

Bob Antol	antol@stargate4173.com; http://www.stargate4173.com
Adam Block	http://www.caelumobservatory.com/resources.shtml
Astro photographer Portraits	http://mstecker.com/pages/app.htm
Tom Boles	http://myweb.tiscali.co.uk/tomboles/index.html
Steve Cannistra	http://www.starrywonders.com/
Russell Croman	http://www.rc-astro.com
Adrian Catterall	http://www.acatterall.com/
R. Jay Gabany	rj2010@comcast.net
Rob Gendler	http://www.robgendlerastropics.com/
Rob Gendler primer	http://www.robgendlerastropics.com/primer.html
Rob Gendler Guide	http://www.robgendlerastropics.com/textcontent.html
Steve Mandel	http://www.galaxyimages.com/UNP1.html
Brad Mead	bradmead@wyoming.com
Jim Misti	http://www.mistisoftware.com/astronomy/
Stan Moore	http://home.earthlink.net/~stanleymm/CCD_topics.html
Sir Patrick Moore	http://www.sirpatrickmoore.com/
Nicolas Outters	http://www.astrosurf.com/nico.outters/astro
Damian Peach	http://www.damianpeach.com/
Jupiter Rotation	http://www.damianpeach.com/barbados06/ jupiter/rotationmovie_thumb.jpg
Philip Perkins	http://www.astrocruise.com/intro.htm
Gordon Rogers	http://www.gordonrogers.co.uk
George Sallit	http://mysite.wanadoo-members.co.uk/sallit/index.htm
Johannes Schedler	http://panther-observatory.com/home.htm
Michael Stecker	http://mstecker.com
Nik Szymanek	http://ccdland.mysite.freeserve.com/index.html

Volker Wendel
Ron Wodaski

www.spiegelteam.de
http://www.newastro.com/wodaski/

Miscellaneous Resources

Abell Catalogue	http://www.icplanetaries.com/abell.html
Apollo Archive	http://www.apolloarchive.com/apidx_apollo_11_b.html
Astronomy Picture of the Day	http://antwrp.gsfc.nasa.gov/apod/archivepix.html
Barnard Dark Objects	http://www.library.gatech.edu/barnard/
Deep-Sky Browser	http://www.deepskybrowser.com/cgi-bin/dsdb/dsb.pl
Digitised Sky Survey	http://www.ngcic.org/dss/dss_ngc.htm
Messier List	http://seds.lpl.arizona.edu/messier/
Minor Planet Checker	http://scully.harvard.edu/~cgi/CheckMP
NGC List	http://www.astrosurf.com/benoit/ngc.html
SOHO	http://sohowww.nascom.nasa.gov/data/realtime-images.html

Equipment Suppliers

Apogee Cameras	http://www.ccd.com/
Astrodon Filters	http://www.astrodon.com/AboutMe.html
CCD Stack	http://www.ccdware.com/downloads/index.cfm
Software Bisque	http://groups.yahoo.com/group/paramount/ ?yguid=82859806
Gradient Terminator	http://www.rc-astro.com/resources/GradientXTerminator/ confirm_download.html
Maxim	http://www.cyanogen.com/
Meade Telescopes and cameras	http://www.meade.com/
Neat Image	http://www.neatimage.com/
RC Optical Systems	http://www.rcopticalsystems.com/contact.html
Registar	http://www.aurigaimaging.com/
SBIG Cameras	http://www.sbig.com/sbwhtmls/newprod.htm
Starlight Xpress Cameras	http://www.starlight-xpress.co.uk/

Weather Forecasts

In my weather favorites folder I have inserted a wedge of climate information sites: you need good information about what is likely to happen "upstairs" before planning your night. I especially favor the series of satellite images because you can draw your own conclusions about how events may unfold. I will refer to Metcheck http://www.metcheck.com/V40/UK/FREE/today.asp?HP189EF, the address for my postcode and to the excellent and comprehensive BBC site at http://www.bbc.co.uk/weather/ukweather/ukpressure.shtml. This gives the opportunity to call for charts of rain, cloud, visible satellite, temperature, wind, and pressure.

Meteorologica Weather Superstore's site:

http://www.meteorologica.info/freedata_lightning.htm gives a rapidly updated look at lightning strikes that might be coming your way and a broader forecast from the same organization is at http://www.meteorologica.info/index.htm

Bob Antol of the Grimaldi Observatory advises me about American weather sites thus:

One of the neatest sites for weather is the Clear Sky Clock. The home page is at: http://cleardarksky.com/csk/ There are over 3,000 clocks in North America. The clock closest to my Stargate observatory is the Thomas J. Boyce Park (Wingdale) clear sky clock, which is approximately 7 miles east northeast of Stargate 4173 at Grimaldi Tower (Poughquag). You can check mine out at:

http://www.stargate4173.com/Phase5/clearSkyClock.htm

From my above page, click on the clock. Then, down the left side, you will see a section "How do I read it?" Most astronomers (at least the amateurs) are aware of the Clear Sky Clocks. They take into account the cloud cover, transparency of the sky, moonlight, etc. Check it out.

There are a couple of other sites that I use, especially when I want to view an animated map of the sky conditions. If you go to

http://www.wunderground.com/US/NY/Poughkeepsie.html?bannertypeclick=suna ndmoon

You can click on the radar image followed by "animate." Of course, there is also "The Weather Channel" Web page at: http://www.weather.com/. Everything is zip code driven. Zip code lookup can be found at:

http://zip4.usps.com/zip4/welcome.jsp

The Stargate is located at zip code 12570. When you enter that zip, you get to:

http://www.weather.com/weather/local/
12570?lswe=12570&lwsa=WeatherLocalUndeclared&from=whatwhere

This is the weather at my location.

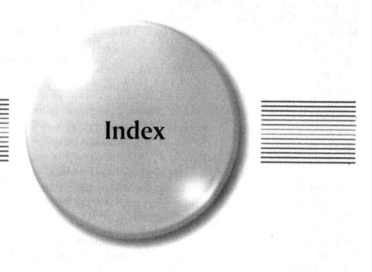

Index